What's Left of Human Nature?

Life and Mind: Philosophical Issues in Biology and Psychology
Kim Sterelny and Robert A. Wilson, Series Editors

What's Left of Human Nature? A Post-Essentialist, Pluralist, and Interactive Account of a Contested Concept, Maria Kronfeldner, 2018

Rock, Bone, and Ruin: An Optimist's Guide to the Historical Sciences, Adrian Currie, 2018

A Mark of the Mental: In Defense of Informational Teleosemantics, Karen Neander, 2017

Mental Time Travel: Episodic Memory and Our Knowledge of the Personal Past, Kourken Michaelian, 2016

Becoming Human: The Ontogenesis, Metaphysics, and Expression of Human Emotionality, Jennifer Greenwood, 2015

The Measure of Madness: Philosophy of Mind and Cognitive Neuropsychiatry, Philip Gerrans, 2014

Beyond Versus: The Struggle to Understand the Interaction *of Nature and Nurture*, James Tabery, 2014

Investigating the Psychological World: Scientific Method in the Behavioral Sciences, Brian D. Haig, 2014

Evolution in Four Dimensions: Genetic, Epigenetic, Behavioral, and Symbolic Variation in the History of Life, revised edition, Eva Jablonka and Marion J. Lamb, 2014

Cooperation and Its Evolution, Kim Sterelny, Richard Joyce, Brett Calcott, and Ben Fraser, editors, 2013

Ingenious Genes: How Gene Regulation Networks Evolve to Control Development, Roger Sansom, 2011

Yuck! The Nature and Moral Significance of Disgust, Daniel Kelly, 2011

Laws, Mind, and Free Will, Steven Horst, 2011

Perplexities of Consciousness, Eric Switzgebel, 2011

Humanity's End: Why We Should Reject Radical Enhancement, Nicholas Agar, 2010

Color Ontology and Color Science, Jonathan Cohen and Mohan Matthen, editors, 2010

The Extended Mind, Richard Menary, editor, 2010

The Native Mind and the Cultural Construction of Nature, Scott Atran and Douglas Medin, 2008

Describing Inner Experience? Proponent Meets Skeptic, Russell T. Hurlburt and Eric Schwitzgebel, 2007

Evolutionary Psychology as Maladapted Psychology, Robert C. Richardson, 2007

The Evolution of Morality, Richard Joyce, 2006

Evolution in Four Dimensions: Genetic, Epigenetic, Behavioral, and Symbolic Variation in the History of Life, Eva Jablonka and Marion J. Lamb, 2005

Molecular Models of Life: Philosophical Papers on Molecular Biology, Sahotra Sarkar, 2005

The Mind Incarnate, Lawrence A. Shapiro, 2004

Organisms and Artifacts: Design in Nature and Elsewhere, Tim Lewens, 2004

Seeing and Visualizing: It's Not What You Think, Zenon W. Pylyshyn, 2003

Evolution and Learning: The Baldwin Effect Reconsidered, Bruce H. Weber and David J. Depew, editors, 2003

The New Phrenology: The Limits of Localizing Cognitive Processes in the Brain, William R. Uttal, 2001

Cycles of Contingency: Developmental Systems and Evolution, Susan Oyama, Paul E. Griffiths, and Russell D. Gray, editors, 2001

Coherence in Thought and Action, Paul Thagard, 2000

What's Left of Human Nature?

A Post-Essentialist, Pluralist, and Interactive Account of a Contested Concept

Maria Kronfeldner

The MIT Press
Cambridge, Massachusetts
London, England

© 2018 Massachusetts Institute of Technology

All rights reserved. No part of this book may be reproduced in any form by any electronic or mechanical means (including photocopying, recording, or information storage and retrieval) without permission in writing from the publisher.

This book was set in ITC Stone Sans Std and ITC Stone Serif Std by Toppan Best-set Premedia Limited.

Library of Congress Cataloging-in-Publication Data

Names: Kronfeldner, Maria E., author.
Title: What's left of human nature? : a post-essentialist, pluralist, and interactive account of a contested concept / Maria Kronfeldner.
Description: Cambridge, MA : MIT Press, 2018. | Series: Life and mind: philosophical issues in biology and psychology | Includes bibliographical references and index.
Identifiers: LCCN 2018001239 | ISBN 9780262038416 (hardcover : alk. paper)
ISBN 9780262549684 (paperback)
Subjects: LCSH: Philosophical anthropology. | Human behavior. | Human beings.
Classification: LCC BD450 .K6993 2018 | DDC 128–dc23 LC record available at https://lccn.loc.gov/2018001239

Tiger Tiger, burning bright, in the forests of the night.
—William Blake

To my father, who knew more about the tiger in us than many

Contents

Preface   xv
Acknowledgments   xxxi

1   Introduction: What's at Issue   1
  1.1   Nature?   1
  1.2   Human?   4
  1.3   Three Different Concepts of Human Nature in Overview   7

I   Three Challenges   13

2   The Dehumanization Challenge   15
  2.1   The Vernacular Concept of Human Nature   16
  2.2   Dehumanization Systematically Viewed   18
  2.3   Social Perspectivity   28
  2.4   The Challenge That Derives from Dehumanization   31

3   The Darwinian Challenge   33
  3.1   What Essences Would Require   34
  3.2   Challenging the Classificatory Role of Essences   41
  3.3   Challenging the Explanatory Role of Essences   49
  3.4   Situating the Anti-Essentialist Consensus   57

4   The Developmentalist Challenge   59
  4.1   From Physis versus Nomos to Nature versus Nurture   60
  4.2   Ignoring Interactions   67
  4.3   The Interactionist Consensus   70
  4.4   What Is the Challenge for a Concept of Human Nature?   85
  Summary of Part I   87

II Three Natures: A Post-Essentialist, Pluralist, and Interactive Reply to the Three Challenges 89

5   Genealogy, the Classificatory Nature, and Channels of Inheritance 91
    5.1 Five Questions Regarding a Species' Nature 92
    5.2 Genealogical Nexus as the Classificatory Nature 96
    5.3 Genealogy and the Channels of Inheritance 102
    5.4 The Resulting Pluralism 114

6   Toward a Descriptive Human Nature 121
    6.1 Descriptive Knowledge about Humans in General 122
    6.2 The Relationship to the Classificatory and the Explanatory Nature 126
    6.3 Typicality Necessary? 131
    6.4 Typicality Sufficient? Or What Does "Important" Mean? 139

7   The Stability of Human Nature 147
    7.1 Innate or Evolved? 148
    7.2 Channelism, Stability, and the Nature–Culture Divide Revived 157
    7.3 A Narrow Enough Concept of Human Nature in the Descriptive Sense 164

8   An Explanatory Nature 169
    8.1 Explanatory Neo-Essentialism 170
    8.2 A Population-Level Solution 179
    8.3 The Explanatory Nature Established 184

9   Causal Selection and How Human Nature Is Thereby Made 189
    9.1 Causal Selection, Control, and Normality 190
    9.2 Choosing among Actual Difference Makers and the Willingness to Control 196
    9.3 How Norms Make Human Nature Visible 202
    9.4 How Norms Make Human Nature Real 206
    Summary of Part II 210

III Normativity, Essential Contestedness, and the Quest for Elimination 213

10 Humanism and Normativity 215
- 10.1 Two Sufficient Entry Conditions for Moral Standing 216
- 10.2 The Ethical Importance of the Descriptive Nature 220
- 10.3 A Dialectic, Essentially Contested Concept of Human Nature 225

11 Should We Eliminate the Language of Human Nature? 231
- 11.1 Elimination versus Revision 232
- 11.2 Redundancy, Neutrality, and Risk of Dehumanization 233
- 11.3 Elimination versus Revision as a Matter of Values 238

Summary of Part III 241

Notes 243
References 265
Index 289

Preface

> Between man and nature hung the veil of culture, and he could see nothing save through this medium. ... Permeating everything was the essence of words: the meanings and values beyond the senses.
> —Leslie White[1]

Human nature has always been a foundational issue for philosophy. It sits like a spider in the center of a web of issues, with threads going in all kinds of directions. Among these threads are the evolution of humans, the nature-nurture or nature-culture divide, the animal-human boundary, innateness, naturalness, genetic determinism, human universals and cultural variation, genetic diversity, enhancement, artificial intelligence and transhumanism, moral standing, speciesism, racism, sexism, normalcy, flourishing, and—last but not least—essentialism.

Central questions directly pertain to the concept of human nature. What does it mean to have a human *nature*? Is the concept the relic of a bygone age? What is the use of such a concept? What are the epistemic and ontological commitments people make when they use the term *human nature*, and do these commitments make sense? These are the questions that drive this book.

In reply to these questions, I will argue for a *post-essentialist, pluralist,* and *interactive* account of human nature. The account is post-essentialist since it eliminates the concept of an essence, which has traditionally been attached to the idea of a human nature; it is pluralist since it defends that there are in the world different things that correspond to three different kinds of post-essentialist concepts of a nature of humans; it is interactive since, first, nature and culture interact at the developmental, epigenetic, and evolutionary levels and, second, since humans repeatedly create parts of their nature via classificatory and explanatory looping effects.

I will establish this account by discussing three epistemic roles of pointing to human nature—classification, description, and explanation—and a few pragmatic functions, most important that pointing to human nature is used for dehumanization, that is, for regarding others as less human. As part of this, I will deal with commitments tied to essentialism, evolution, heredity, the nature-nurture divide, and normalcy assumptions. Since the mentioned commitments transgress the boundaries between philosophy, sciences, and politics, this book has to do so too. The core of the book is nonetheless developed from a philosophy of science point of view but one that reaches out to social, moral, and political philosophy.

Two Ways of Discussing Human Nature: An Empirical and an Analytic-Reflective Stance

The book's scope is limited to one of two ways of discussing human nature: discussing *human* nature and discussing human *nature*. This book concerns the second issue, which is an analytic and reflective issue about what "having a nature" and "something being due to nature" mean.

The first issue, by contrast, is about what it means to be human rather than another animal and therefore about which traits belong to the human way of being. With respect to that, a lot of empirical disagreement reigns. While some (e.g., Lorenz 1960) have claimed that our nature is particularly evil since we are unique in killing our own kind on a massive scale, others have defended that it is in our nature to be altruistic and cooperative (e.g., famously, Kropotkin 1902). Such disagreements are not settled, as Pinker (2011) testifies. Still others focus on our upright gait, or tool use, or the opposable thumb as what makes (or made) us human. They will be opposed by still others, especially from philosophy, who believe that what characterizes humans cannot be any physiological or anatomic character. Only phenomena that go beyond what sciences can grasp—reason, mind, morality, consciousness, creativity, or the like—will do the trick, or so the argument goes. Some might even extend such a line of reasoning toward the claim that artificial intelligence and enhancement will basically overcome the current physiological and anatomical limits for the respective bodily or cognitive capacities so that the future will be posthuman or transhuman, often thereby asserting that intelligence or some other purely mental property is the essence of what it means to be human. All of these claims are empirical, and disagreement regarding them concerns which properties—as a matter of fact—make up human nature rather than the nature of any other animal, or, for that matter, the nature of machines. To ask about human nature would thus be doing empirical sciences, understood in a way that includes philosophical anthropology in the sense of a generalized version of seeking an empirical understanding of how humans are. Most philosophical books that have the term *human nature* in the title are concerned with such empirical issues. The particular characteristics of the human life form that are discussed (e.g., altruism and cooperation versus egoism and aggression) are, in the frame used here, the possible content of the concept of human nature. At issue is what it is like to be a human, that is, *how* we are.[2]

The second issue, the analytic-reflective issue, which is this book's sole concern, is about what it philosophically means that there is an alleged human nature. The book is thus not concerned with the possible *content* of a concept of human nature (i.e., which properties belong to an alleged human nature); it is, rather, concerned with the *ontological status, epistemic roles*, and *pragmatic functions* of an understanding of being human that refers to a "nature." Ontological status is about whether human nature exists, that is, whether the concept refers to something real, out there in the world, existing independent of our minds. It concerns, so to speak, the metaphysics of human nature. Epistemic roles specify what those people who produce knowledge about humans use the concept of human nature for—for classification, description, or explanation. It concerns the epistemology of human nature. Pragmatic functions disclose why the epistemic roles are important for those interested, that is, why one cares for knowledge about human nature. It targets the pragmatics of human nature and specifies why human nature is important. To discuss the status, roles, and functions of a concept—to discuss what it means to refer to a human nature—is, methodologically, not an empirical endeavor; it is a reflective philosophical exploration: it analyzes and reflects on the way one "sees" things in the sense of White's (1958) quote above. It reflects on *how one sees things through one's concepts, theories, assumptions*, and so on. In the case of this book, the reflection is on how one sees humans through the lens of a concept of human nature. This book is thus an initiative in a reflective philosophy of human nature that combines metaphysical, epistemic, and pragmatic considerations.

The book is as neutral as possible with respect to the content issue, that is, empirical questions about this or that property being part of our nature.

Keeping the book neutral in that sense is essential since failing to do so would make the book too dependent on the shifting empirical knowledge on how we are, and thus too dependent on the continuing disagreements about the relevant empirical knowledge—disagreements among philosophers as well as between philosophers and others, such as sociologists, cultural anthropologists, primatologists, archaeologists, and biologists. This book aims at those ontological, epistemological, and pragmatic issues that do not change with the empirical knowledge or disciplinary focus and that are in that sense more fundamental. Yet I hope that readers who are primarily interested in the content issue will nonetheless find this book interesting. After all, an interest in the content issue might well presuppose opinions on the reflective issues about the status, epistemic roles, and pragmatic functions of talking about humans as having a nature.

Four Building Blocks, Three Challenges, and a Constructive-Revisionary Core with an Eliminative Rhetorical Finish

The book assumes four core traditional aspects—building blocks—of concepts of human nature: *specificity, typicality, fixity,* and *normalcy*. Talk about human nature is, after all, often taken to refer to properties that are specific and typical of the species, that involve some fixity and that constitute normalcy, or how a member of the species should be.

The ultimate aim of the book is revisionary: to make sense out of these aspects, given three contemporary challenges that the concept of human nature faces. I address these challenges in part I. There is (what I call) the "dehumanization challenge," what R. A. Richards (2010) has called the "Darwinian challenge," and what Schaffner (1998) has called the "developmentalist challenge."

As I understand it, the dehumanization challenge addresses the worry that in social affairs, the concept of human nature is purely functional: in different contexts filled with different content, only the function(s) stay the same. One such function is dehumanization, which is used by those who note what they have and others lack, so that in each case, the respective others can be dehumanized—regarded as less or not human. The vernacular concept of human nature is thus devoid of any fixed content and characterized by a dehumanizing social perspectivity. With this, dehumanization threatens the concept's ontological status since it takes away the concept's objectivity. Dehumanization also illustrates how normalcy enters talk about human nature: those who are normal are more human than those who are regarded as abnormal. With respect to the core four building blocks, the dehumanization challenge mainly concerns typicality and normalcy assumptions.

The Darwinian challenge is of a completely different sort, but it also threatens the concept's ontological status. According to the classificatory and explanatory schema of Darwinian ontology, there is simply no human nature in the sense of an essence as traditionally conceived. Since such essences played an epistemic role simultaneously for classification, explanation, and description, the Darwinian challenge tackles not only the ontological status of human nature but also its epistemic importance. Since it concerns all four of the building blocks—specificity, typicality, fixity, and normalcy—the Darwinian challenge is the deepest of the three challenges at issue in this book.

The developmentalist challenge is mainly about fixity and concerns the interaction of nature, culture, and environmental factors—an interaction that at its core is developmental (since it happens at the level of individuals), even if it also has relevance for an intergenerational and an evolutionary level. The contested issue is whether one can and should distinguish human nature—in its explanatory role—from nurture (i.e., cultural and environmental developmental resources), which might be equally explanatory for the human life form. The developmentalist challenge is easier to overcome than the Darwinian challenge since it is less connected to essentialism. Yet overcoming it involves finding a new and more interactive interpretation of the idea of fixity, which usually comes with the concept of human nature and has often been mounted on innateness assumptions. Many consider the concept of human nature to be obsolete because they cannot envision such an interactive account of the fixity aspect. It is one of the major contributions of this book to try to overcome this obstacle.

In a nutshell, dehumanization challenges the objectivity in any concept of human nature; Darwinian ontology challenges the traditional essentialism in the concept of human nature; and developmental interactionism challenges the often assumed contrast between nature and other developmental resources, such as culture, society, and environmental factors, usually summarized under the label "nurture."

These three challenges against human nature, and the implied nature–nurture divide, will be the foil to arrive at a post-essentialist, pluralist, and interactive account of human nature in the sciences. The account is revisionary and relies on a matrix of different concepts of human nature—different ways of saying something salient and objective about being human—despite the three challenges. The concept of human nature survives the three challenges, but not as one concept: it survives as a tripartite combination of a classificatory, a descriptive, and an explanatory nature of being human. Rather than throwing out the baby (the concept of human nature) with the essentialist bathwater, I have tried in this book to traverse the middle ground by putting an age-old strategy to epistemic use: to divide (take the epistemic roles apart that the concept played so far) and conquer thereby (find that there is something to rescue, despite the challenges that the concept faces). As part of establishing this plurality of post-essentialist human natures, I will also untie the tangle between nature and culture. This will lead to a defense of nature and culture as distinct channels of inheritance.

Part III deals with three issues that concern further pragmatic functions connected to normalcy, the fourth building block of concepts of human nature. I will describe how my account can integrate a normative use of the concept of human nature, a use that belongs to moral and political philosophy. I will also discuss why it is unlikely that there will ever be a consensus on the content of human nature. For this, I will point to what political scientists have called *essential contestedness*. By continuing to discuss concepts like democracy, truth, or human nature, the negotiating community comes closer to what is generally meant by the respective concepts, independent of or even because of the plurality in meaning. Human nature, according to the account developed here, is such an essentially contested concept. It would be missing the point of the concept if one eliminated the plurality in how the concept is used epistemically and filled with content.

Nevertheless, it might be advisable to eliminate the term *human nature*. After all, the pluralist replacements of the essentialist concept could all completely abstain from the language of human nature. The discussion of this rhetorical dimension will draw on a panoply of pragmatic functions of the expression *human nature* and conclude that we should avoid the term as much as possible. It is a term that is like a Wittgensteinian ladder: it was historically useful to get to where we are, but as part of a post-essentialist and pluralistic account of concepts of human nature, it is time to throw the term *human nature* away.

Situating the Book within Current Literature
The book covers a debate in philosophy of science that started roughly in the 1980s and has gained momentum recently. Given the book's breadth in taking three important challenges simultaneously into account and given that it distinguishes five distinct analytic layers (content, ontological status, epistemic roles, pragmatic functions, and essential contestedness), with a focus on the last four layers, this book is (to the best of my knowledge) the first of its kind, even though the literature on human nature that accumulated since the mid-1980s is huge.

A couple of edited collections provide good overviews of the variety of issues connected with human nature that I mentioned at the beginning (e.g., Roughley 2000; Inglis, Bone, and Wilkie 2005; Sandis and Cain 2012; Downes and Machery 2013). There are also encyclopedia-style introductory articles on the concept of human nature with various foci (e.g., Shapiro 1998; Stenmark 2012; Kronfeldner, Roughley, and Toepfer 2014). There are seminal philosophical papers on the topic (e.g., Sober 1980; Hull 1986). There are companions for the classrooms (e.g., Trigg 1999; Kupperman 2010; Stevenson, Haberman, and Wright 2012), usually comparing different classical philosophical positions on human nature, from Confucius to sociobiology, mainly addressing the diversity of content with which the concept can be filled (for an overview, see Jaggar and Struhl 2014). Then there are very specific contributions. Bayertz (2013), for example, traces the history of how philosophers classically addressed the issue of human nature by focusing on which role humans' upright gait played in philosophy. Some books consider human nature as the foundational concept for philosophy in general, contributing to the kind of empirically grounded philosophical anthropology mentioned above (e.g., Hacker 2010). There are also collections that focus on specific analytic issues, for instance, whether enhancement conflicts with the fixity of human nature (e.g., Sandel 2007; Buchanan 2011; Özmen 2011).

More oriented to topics dealt with in this book are books on human nature that deal with evolution, heredity, and development. Among these, some defend a specific version of the concept of human nature, while others show that culture is more important than nature to explain how humans are. Dupré (2001), Buller (2005), and Sahlins (2008), for instance, all argue against evolutionary psychology's concept of human nature, with Dupré and Sahlins adding arguments against the idea of humans as egoistic. There are books that address the developmentalist challenge and whether human nature exists: Pinker (2002) argues for nature and Ridley (2003) and Prinz (2012) for nurture. Some argue, as this book does, for a decisively interactionist picture (e.g., Richerson and Boyd 2005). There are also books that are closest to this one in analytic orientation but more focused on one of the challenges at issue here (e.g., Dupré 2001; Keller 2010; Tabery 2014; Schaffner 2016; Mikkola 2016; Lewens 2015; Milam (forthcoming). The goal in these books is not, at least not predominantly, to take sides—in the sense of claiming that human nature exists or not, that culture is more important or not in explaining how humans are. The goal in these books, as in this one, is analytic: to analyze the historical or philosophical assumptions structuring the respective debates. Dupré, Keller, Tabery, Schaffner, and Lewens discuss the developmentalist challenge, though each has a different focus; Mikkola discusses the dehumanization challenge and Milam the historical settings (such as the Sputnik shock, World War II, or the civil rights movement) that went into North American discussions about humans as by nature violent or not. Yet I do not know of a book with an analytic orientation that covers all of the three contemporary main challenges together—the dehumanization challenge, the Darwinian challenge, and the developmentalist challenge—and gives a constructive, balanced, and at the same time critical philosophical reply.

So far the analytic discussion in focus here has mainly been dealt with in a series of recent papers, all taking Hull's (1986) paper as an anchor. Within that body of literature, ontological status has been regularly addressed since Hull (1986) launched his attack against the concept of human nature. The diversity of epistemic roles connected with the concept has been explicitly addressed since only recently, in, for example, Roughley (2011), Samuels (2012), Kronfeldner et al. (2014), and Machery (2016a). Pragmatic functions are mentioned occasionally, as in Antony (1998, 2000), but they have not been analyzed systematically and in detail in connection with the other two challenges. Essential contestedness (to the best of my knowledge) has so far not been addressed with respect to human nature.

Given the state of the debate (exemplified most recently by Hannon and Lewens's forthcoming collection), only an in-depth coverage in book length can prevent the extremes that dominate popular side-taking accounts as well as paper-length analytic interventions in the debate. A bigger and constructive picture is needed—an account that charts the midway between scientifically outdated thinking about human nature, on the one hand, and throwing out the baby with the essentialist and dualist bathwater, on the other hand, an account that divides things up analytically and conquers thereby.

Chapter-by-Chapter Outline

Chapter 1 starts with some basic assumptions that concern the meanings of the term *nature* and the four aspects that reoccur with respect to concepts of human nature: specificity, typicality, fixity, and normalcy. On that basis, a family resemblance of different concepts of human nature is postulated. Then the connection to and focus on issues from the philosophy of the life sciences is introduced by justifying that the term *human* refers to either a biological group category (e.g., *Homo sapiens*), labeled as "humankind," or to a socially specified group, labeled as "humanity." The two groups—humankind and humanity—are presented as distinct but overlapping. It is made clear that the book mainly concerns humankind and that humanity will be addressed only as part of the dehumanization challenge and in part III. Given the assumptions made and given the distinction of the three different epistemic roles (classification, description, and explanation), I outline how the account will ultimately defend that there are—for each kind of group—three natures: a classificatory nature, a descriptive nature, and an explanatory nature. The pluralism in this rests on the claim that different disciplines use a different epistemic role and thus a different concept to say something salient and objective about humans. The chapter ends with a brief justification for why a fourth, a normative, concept of a human nature is decoupled from the post-essentialist, pluralist, and interactive scientific picture that will be developed in part II.

Part I analyzes the three standard challenges that the concept of a human nature faces in contemporary debates in science, philosophy, and politics.

Chapter 2 is on the dehumanization challenge that results from use of the concept of human nature in social affairs for regarding some people as less human. By introducing dehumanization in a systematic manner, illustrated by examples, mainly with respect to sexism and racism, the aim of this chapter is to show that the folk or vernacular concept of human nature is purely functional and perspectival: the content is exchangeable and varies with the perspective of those who speak, whereas the function (usage for dehumanization) remains the same across contexts. Based on this, the dehumanization challenge is characterized as twofold: from the scientific point of view, the social perspectivity in the vernacular concept needs to be overcome to arrive at an objective concept of human nature; from the social point of view, dehumanization itself is the challenge: it needs to be overcome because it conflicts with fundamental ideas about equality, human rights and justice.

Chapter 3, on the Darwinian challenge, deals with essentialism as challenged from within science. Within science, the nature of a species has traditionally been conceptualized as an essence. After an introduction of traditional essences as simultaneously playing a classificatory, descriptive, and explanatory epistemic role, the connection to the concept of natural kinds and the species category is introduced. On that basis, the chapter reconstructs the anti-essentialist consensus in contemporary philosophy of the life sciences, a consensus that is based on the consequences of the Darwinian ontology. By discussing in detail the classificatory and the explanatory role of an essence, it will be shown that given Darwinian ontology, there are no necessary and sufficient conditions for membership in a biological species (fulfilling the classificatory role of an essence) that are at the same time (as required from an essence) fulfilling an explanatory role for the traits that are characteristic of the kind (descriptive role). This is the core of the Darwinian challenge, leading to the question: What is left of human nature if essences of biological species are eradicated from the contemporary biological ontology?

The developmentalist challenge, addressed in chapter 4, stems from critiques of dualistic pictures of ontogenetic development. At issue are attitudes that regard phenotypic traits as "due to nature" or "due to nurture." The chapter introduces the so-called nature–nurture divide by comparing its meaning and role in Greek antiquity and during the hardening of the divide in the nineteenth century, which is presented as depending on Francis Galton's anti-Lamarckism. Part of the hardening was the assumption that the causality between nature and nurture (or nature, culture, and environment) can be apportioned so that traits can be said to be due to either nature or nurture. That such an apportioning is impossible is part of the interactionist consensus, which is presented as also opposing genetic determinism or other forms of giving priority to nature over nurture, be it at the level of development (which is regarded as core level), epigenetic inheritance, or comparing biological and cultural evolution. The chapter ends with a discussion of what is still controversial about the interaction of nature and nurture, to better see what exactly results from it as a challenge for the concept of a human nature.

Part II rescues as much as possible from the battlefield. It is the core constructive part of the book. It assumes that sciences can get rid of the social perspectivity discussed as part of the dehumanization challenge. On that basis, it revisits the Darwinian and the developmentalist challenge and develops from these a post-essentialist, pluralist, and interactive account of different concepts of human nature in the sciences. Each of these post-essentialist human nature concepts refers to a state of affairs (something existing in the world) that fulfills only one epistemic role (classification, or description, or explanation), whereas the essentialist account had one thing (the essence) fulfilling all of these epistemic roles simultaneously.

Chapter 5 discusses genealogy (ancestor-descendant relationships) as the post-essentialist classificatory nature of humans and the channels of inheritance that result from it. It starts with five distinct questions regarding the nature of a species and defends that, given contemporary species concepts in evolutionary thinking, genealogy is relevant for all five questions, even though for some only indirectly, nonspecifically, and partially. It is then defended that genealogy can ground the classificatory nature of biological species. Finally, it is shown that there are dynamically different channels of inheritance and that the biological channel conveys much more stability for the reoccurrence of developmental factors than the cultural channel. Combining these points, the general claim is that genealogy not only grounds the classificatory nature of biological species; it is also indirectly, partly, and nonspecifically relevant for explaining the respective life form of a species because it is at the foundation of that very channel of inheritance that guarantees a high stability of developmental resources. After that, the pluralism of the three natures is introduced. The chapter ends with a historical and social contextualization of the importance of genealogy.

Chapter 6 builds on this and takes the first steps toward reestablishing a descriptive human nature. After discussing why sciences need a concept of a descriptive nature, the latter is decoupled from the classificatory and the explanatory nature of humans. Given that even the severest critics of the concept of human nature leave a version of the descriptive concept alive, disagreements in the current literature about the details of how exactly to specify it, in particular with respect to typicality, are addressed. After specifying what typicality means, the chapter addresses whether typicality is necessary, given that species show polymorphisms (such as sex-specific traits). Are polymorphisms part of the nature of a species? This question leads to an important issue about abstraction, since—as I will argue—almost any trait can be made typical with abstraction. Finally, various further qualifiers are discussed since typicality alone turns out to be insufficient to have a descriptive concept of human nature that is narrow enough to exclude typical cultural habits from being part of human nature, such as "to carry cell phones" or "to bury the dead." What is added to typicality, however, is presented as depending on disciplinary interest or similar perspectives. Yet fixity of traits is a frequently recurring qualifier.

Chapter 7 then reconstructs fixity of human nature as stability. First, fixity as innateness is dismissed since it neither helps to solve the developmentalist challenge nor is it a concept that includes an evolutionary level of description and explanation. The next proposal that is reviewed is one that claims that one needs to add evolvedness (rather than innateness) to typicality: if a trait is typical and evolved, it is part of human nature. Although this proposal, mainly defended by Machery (2008, 2016a, forthcoming), is on the right track, it is also dismissed since it is still too broad, mainly because culture can evolve too. What is valuable about the proposal is that it points to the temporal dimension of a species' descriptive nature. The claim that is defended on the basis of this is that populations of individual organisms show typicality (similarity) not only in space (synchronically) but also over time (diachronically). The typicality over time is what I call stability. Given that the biological channel of inheritance (as introduced in chapter 5) reliably provides a high stability for the reoccurrence of developmental factors, human nature in the descriptive sense is reconstructed as traits that are typical and reliably reoccur over time because of developmental resources that travel the channel of biological inheritance. Since this seems to directly fly in the face of the interactionist consensus, it is discussed how it is possible to claim that a trait is in that sense "due to nature" without contradicting the interactionist consensus. Taking abstraction seriously and distinguishing between difference-making explanation and production explanation is a central part of the reply.

Any descriptive nature has an explanatory counterpart, which is addressed in chapter 8. Human nature in the explanatory sense refers to those developmental resources that are typical and biologically inherited. To defend that claim, two revisionary essentialist proposals—dubbed developmental essentialism and teleological essentialism—are reviewed and dismissed. These proposals claim to rescue an essentialist explanatory concept of human nature. The aim of these "new" essentialisms is to prevent the problems of the traditional essentialism by using the explanatory role of essences only. My claim against these essentialist proposals is that both suffer from an intrinsicality bias. They are too dependent on the assumption that an explanatory nature needs to be internal to individual organisms. The suggestion is then to reconceptualize the explanatory nature as being internal to a population rather than to the individual organisms. The explanatory nature of humans is then specified as a population-level pool of developmental resources that travel the biological channel of inheritance. This populational explanatory nature is perfectly compatible with the demands that come from the Darwinian as well as the developmentalist challenge.

Chapter 9 deals with causal selection and how human nature is made thereby. Explaining human traits involves, as all other explanations do, an epistemic choice: some causal factors are ignored, while others are selected to be included in an explanation. The chapter introduces an approach to causal selection that shows how normative stances (preferences and values) make causes first visible and then real: how our normative stances and values bias us toward certain kinds of explanations (e.g., explanations pointing to human nature) and how such explanatory endeavors—via explanatory looping effects—make the selected causes real. That way, the chapter can show that human nature in the explanatory and descriptive sense is not only interactive in the sense that nature and culture interact at the developmental, intergenerational and evolutionary level. Human nature is also interactive in the sense of explanatory looping effects: when one explains human life by pointing to human nature, this can influence humans in their behavior; depending on whether people move in or out of the explanation, this stabilizes or changes human nature over time. This is how humans in part make their nature.

Part III deals with normativity, essential contestedness, and the quest for linguistic elimination. The scientific picture of part II leaves out the normativity that is traditionally attached to the vernacular concept of human nature. It also ignores that the vernacular concept often refers to social rather than biological groups and also what follows from the dehumanization challenge for a scientific picture as established in part II. These three topics are dealt with in the final part III.

Chapter 10 is on humanism and normativity. Humanism is introduced as a view that insists that all humans are equal, subject to human rights and norms of justice. This clearly involves normative evaluations about how humans should live, that is, how they flourish. Where has all that normativity gone to, if it is not anymore part of the post-essentialist, pluralist, and interactive account of human nature in the scientific, objective sense? After discussing the issue of moral standing (who belongs to the group of individuals who count for certain moral considerations), the chapter discusses how the descriptive nature (as reconstructed in this book) can be of ethical, moral, and political importance to determine what is needed for human flourishing. This is then combined with the results of chapter 9, leading to a dialectic concept of an ever-changing descriptive and explanatory looping human nature. Finally, this dialectic concept is interpreted as an essentially contested concept: agreement on the content of an essentially contested concept is unlikely since it is essential to the concept to be contested.

Finally, chapter 11 asks whether we should eliminate the term *human nature*. The pluralism defended in part II means that the term has become ambiguous and redundant for describing the matters of fact that scientists want to describe with it. In addition, the risk of social harm following from using the term *human nature* is high, given the insights on dehumanization from chapter 2. Even the post-essentialist human natures (the descriptive, explanatory, and classificatory nature) can lead to dehumanization. On the basis of this and the insights from parts I and II, the chapter shows three balancing problems if one wants to decide whether the term *human nature* should be prevented, given the post-essentialist, pluralist, and interactive account. These balancing problems involve epistemic values that relate to the epistemic roles of the concept, as well as social values that relate to the pragmatic functions of the concept. Taken together, they lead to important trade-offs of values and consequences that nonetheless allow the application of a precautionary principle. This then, finally, shifts the balance toward a regulative ideal that prescribes eliminating the language of human nature. The concepts of human nature, I thus conclude, can stay, but the old terminology should be prevented as much as possible.

In a Nutshell

This book starts with a couple of assumptions to narrow the scope, addresses three major challenges in part I, defends that there are three post-essentialist natures in part II, and addresses a couple of normative issues in part III. In tone, it starts critically, goes constructive-revisionary, and ends with an eliminative finish.

Acknowledgments

My deepest thanks go to Alexander Reutlinger, who has accompanied and influenced this book over the years in many ways. Even though human nature is not among his own philosophical interests, his impact shines through many of the thoughts that found their way into this book. I also owe a lot to my coauthors of the article "Recent Works on Human Nature," Neil Roughley and Georg Toepfer. I first considered and defended the idea for the kind of post-essentialist pluralism that I develop in this book with them. They in particular, but also Peter McLaughlin and Markus Wild, the other two members of the working group on human/animals of the German Research Foundation Network for Philosophy of the Life Sciences (DFG fund KR3392/2-1), indirectly influenced the book's current shape, given the open and in-depth discussions that only such a network setting can offer. My thanks go not only to them, but also to the other members of the network and to the DFG for funding the network. Yet the most crucial and chief financial and institutional support for the book project itself came from one person, Martin Carrier, whose generosity will forever be unmatched, a generosity that allowed me to enjoy a two-year phase of freedom—to travel and research, without departmental and teaching duties—first at the Center for Philosophy of Science at the University of Pittsburgh, then at the Fishbein Center for the History of Science and Medicine at the University of Chicago, and, finally, at the Max Planck Institute for the History of Science in Berlin. I also thank these three centers for their energizing environment and support. In particular, I thank Edouard Machery, Sandra Mitchell, Kenneth Schaffner, Robert J. Richards, and Veronika Lipphardt. During that time of freedom, they were my Poincaréian hooks that kept my ideas from straying too far. At the very beginning of this project, however, it was Paul Griffiths and Karola Stotz, and the hospitality of the Sydney Center for the Foundations of Science that helped this project get off the ground. I thank them and the others I have already mentioned for the opportunities, interactions, and discussions. Early drafts of individual chapters were read by the participants of the Human Nature Workshop at Chicago (headed by Robert J. Richards) and my students of the 2015 Human Nature Course at Central European University; the first draft of the complete manuscript was read and commented on in print by Mara-Daria Cojocaru, Jürgen Kufner, Jessica Laimann, Alexander Reutlinger, and three anonymous referees for MIT Press. I benefited a lot from their remarks. Thanks also go to Mark Brown, Friederike Eyssel, Evelyn Fox Keller, Lisa Gannett, Wolfgang Gebhardt, Lily Huang, Ferenc Huoranszki, Philip Kitcher, Richard Lewontin, Francesca Merlin, Diane Paul, Olivier Rieppel, Hans-Jörg Rheinberger, Marshall Sahlins, Isabella Sarto-Jackson, William Sterner, Robert A. Wilson, and William C. Wimsatt for the helpful discussions on specific aspects or chapters of the book. Olesya Bondarenko deserves thanks for assisting me in obtaining permissions and indexing the book, through which she also indirectly helped in the final streamlining of the words used. I thank Vera Brüggemann for bringing the organic, the abstract, and the graphic so nicely together in the figures she prepared for the book. Finally, I thank the series editors, Kim Sterelny and Robert A. Wilson, as well as Philip Laughlin, Anne-Marie Bono, and Judith Feldmann from the MIT Press, and Beverly Miller, for making the publication process of this book such a gratifying experience.

Some chapters contain material in changed form that I have published previously. I thank my coauthors and the publishers for the permission to reuse the material. Details and credits on such reuse can be found at the beginning of chapters 2, 3, 4, and 9, in the first footnote related to these chapters. Credits and the permissions to reprint previously published figures from other authors are included in the respective figure captions.

1 Introduction: What's at Issue

This chapter establishes anchors for the discussion by making some important assumptions explicit. These concern the meaning of the term *nature*, four aspects of concepts of human nature, and the reference to *Homo sapiens*. I then introduce a post-essentialist, pluralist, and interactive account that I defend in part II. The core of that new account of human nature is that it encompasses three different concepts of having a human nature: a classificatory human nature, a descriptive human nature, and an explanatory human nature. This introductory chapter therefore sets the frame for the rest of the chapters that go into the necessary details of specific issues regarding this new account.

I begin in section 1.1 by introducing two core meanings of *nature* and four reoccurring aspects (building blocks) of concepts of human nature: specificity, typicality, fixity, and normalcy. In section 1.2, I establish that the term *human* can refer to either a biological group (e.g., the species *H. sapiens*) or a socially delineated group. These two groups are shown to overlap even though they are distinct. In section 1.3, I introduce the three natures: classificatory nature, descriptive nature and explanatory nature. I also justify why normalcy is not going to be part of the account developed in part II.

1.1 Nature?

Natures of Things and Things of Nature
There is a mind-boggling variety of meanings attached to the term *nature*, which derives from the Latin *natura*, ultimately going back to the Greek *physis*. As R. Williams once said ([1976] 2011, 186), "Any full history of the uses of nature would be a history of a large part of human thought." I

cannot provide any such full history or even of a broad systematic account of the diversity of meanings of nature.[1] All I will do is distinguish between two meanings that are quite basic and show up in two typical locutions: that there are *natures of things* and *things of nature*.[2]

The first—that there are natures of things—has traditionally been taken to refer to the essences of kinds, which etymologically has strong connections to growth and biological reproduction. This meaning is of central concern in this book. Nature in the second sense, referring to a domain of things that are investigable in a systematical manner oriented toward accessible evidence about them, is another very old, basic meaning. The rainbow and other so-called natural phenomena were the first things in the history of philosophy in the West (i.e., the tradition beginning with Greek antiquity) that became naturalized in that sense. They were not any more conceived as mythological but as natural.[3] This meaning of *nature* will mostly be in the background of this book. It sets the frame for this book and for any search of general empirical knowledge about humans. David Hume called that kind of knowledge in his *Treatise of Human Nature* an "accurate anatomy of human nature" (Hume [1739] 1896, 263). Today one might simply consider it the knowledge produced in human sciences—all those sciences (in the broad sense) concerned with humans: human biology, social sciences, and the humanities.

What unites the two basic meanings is that both are used in a dualistic (i.e., antithetical, contrastive) manner. Since Greek antiquity, nature (in both basic senses) has been conceived as carrying with it a contrast: natural versus supranatural and natural versus cultural, to name only two of the contrasts that form a dualistic landscape around "nature." Therefore, if something is part of human nature, this can imply that it is not supranatural, not due to nurture or culture, not superficial, not artificial, and so on. The concept of nature thus varies with its contrast. This contrastive character of the concept of nature is important for two reasons.

First, the ontological attitude of the book is naturalistic. It thus assumes the contrast supranatural/natural as a background. As part of the naturalistic outlook, I start with ontologically assuming that there are individual organisms (particulars), properties (of organisms, understood to be non-relational, i.e., intrinsic and qualitative), relations (between individual organisms or properties), and kinds (group categories). It is also assumed that organisms are real (i.e., mind-independently existing entities) and that one can get sufficient epistemic access to them, especially via the sciences, to legitimately claim that one has knowledge about them. This entails neither any stance regarding the (non-)existence of minds (they might exist or not), nor any stance regarding the relationship of minds to organisms. The critical topic of the book is thus not whether a naturalistic account of humans is possible, but whether there is, given a naturalistic outlook, in addition to the organisms with bodies and minds, properties, relations, kinds, a "nature" of those very organisms that are conventionally called by the name *human*. Part of this is to ask whether the only option for admitting that there is such a "nature" is admitting that there are essences of biological species.[4]

Second, the contrastive nature signals that the concept is used for creating boundaries. As a consequence, some group categories incorporate the nature-culture contrast—for example, animal/human, machine/human, race/ethnicity. These in turn point to a bunch of different meanings of the nature/culture divide: nature/reason, biology/culture, innate/acquired, heredity/development, genes/environment, sex/gender, and so on. These meanings are connected to certain dualistic connotations: prenatal/postnatal, fixed/malleable, given/made, essential/accidental, ultimate/proximate, universal/local, internal/external, normal/pathological, and more. Finally, in addition to these contrasts (often stabilizing each other), there are other bipolar distinctions that support these contrasts. Probably most important today are science/society and science/technology; probably most important in the nineteenth century were mind/body, civilized/uncivilized and, further down the road of history, physis/nomos at the beginning of philosophy in the West.[5] The landscape of dualisms around the concept of a human nature indicates that it might not be easy to get rid of such a concept since it is so entrenched. It also indicates that the concept has something to do with dividing things—properties, causes, and people, in a scientific sense but also, as chapter 2 will show, in social affairs.

Four Aspects

Building on Griffiths (2002) and Roughley (2011) but also going beyond them, I assume that there are four aspects—building blocks—of the various concepts of human nature: *specificity, typicality, fixity*, and *normalcy of traits*. Talk about human nature is, after all, often taken to refer to properties that are specific and typical of the species, involve some fixity, and constitute normalcy, that is, how a member of the species should be. If all four aspects are part of the concept, then a trait (e.g., rationality) is part of human nature if this trait is specific and typical of humans, somehow fixed in human beings, and something a normal human being should exhibit.

A concept of human nature can involve all of these aspects, a few, or just one. In Wittgensteinian terms, this means that there is a family resemblance between different concepts of human nature.

The first and second aspects raise the question of essentialism and kind thinking, the third raises the question of heredity and the nature-nurture divide, and the fourth raises the question of whether and how normalcy should be part of understanding the human in science, politics, ethics, or morality.

1.2 Human?

Homo sapiens as Humankind

The term *human* often refers to the biological species *H. sapiens*; I call the group thus delineated *humankind*. Humankind is the group that is in the background of discussions about human nature in philosophy of science, which started, roughly in the 1980s, when Hull (1986) launched a forceful attack against it. It is also in the center of this study. If there is humankind, then there need to be criteria for organisms partaking in it. Criteria that determine the boundaries of a kind, and thus membership in it, will be called *classificatory nature*. The *genealogical nexus* between *H. sapiens*, a relation between organisms, will be discussed as a candidate for such a classificatory nature of humankind, determining the boundaries of and membership in the species *H. sapiens*.

I will henceforth talk about *H. sapiens* or humankind if the species and related concepts of human nature are at issue, except in cases where the context already makes this meaning of "human" transparent. I will ignore the possible extension of "human" to the genus *Homo*, to a tribe, a subfamily, or even, as once, the family Hominidae. The history and the latest changes in how the term *human* is integrated into biological terminology shows that the reference is shifting constantly.[6] For this study the important point is that wherever one stops on this slippery slope, it would be an extension in the same dimensionality of the concept, namely, that the term refers to a biological group. Since not much depends for this study whether *human* refers to *H. sapiens* or a larger biological group, I will leave it at that.

Humanity

It is, however, of utmost importance that the term *human* can also refer to a purely social category and thus be applied with a different dimensionality in mind. I use the term *humanity* for the respective social group. If used for a social group, "human" first and foremost refers to us rather than them (whoever that is). Membership in such a social group is determined by a set of traits that this group at a certain time regards self-referentially and discursively as necessary and sufficient for being a person and thus a member of humanity. If it refers to biological issues (e.g., being a member of *H. sapiens*), it does so only contingently, that is, only because those speaking at a certain point in history have accepted themselves as being *H. sapiens*.

Hull (1986) also mentioned this social category (humanity) but decided (and explicitly so) to ignore it. I cast the net a bit wider but still keep a focus on "human" as referring to a biological group category. Doing so allows me to study a specific challenge in more detail than Hull did, namely, the dehumanization challenge, and it allows me to show that humanity also (as humankind) has a nature, or natures, to be precise. I analyze the reference to social groups in detail in chapter 2, under the label "dehumanization," in order to identify which challenge it provides. I will ultimately suggest that "humanity" is also referring to a moral group. The reason is that only humanity, not humankind, can function as foundation for moral and political philosophy, as in, for example, Rawls's (1971) theory of justice. Philosophers will connect humanity with personhood, rather than with specieshood, and rightfully so. Yet that group also has a classificatory nature, a descriptive nature, and an explanatory nature (an issue discussed in part III). After humanity takes center stage in chapter 2, it will be in the background in chapters 3 and 4 and part II. That is how I hope to cast the net wider than Hull (and others since him) to transgress the boundary toward politics and moral issues, while still being focused on humankind and the ontological and epistemological issues that relate to talk about human nature.

If both dimensions, the biological and the social, are taken into account for the overall picture, a pluralist matrix can be derived, where "human" can refer to either a biological group or a social group, each with three respective natures attached: a classificatory, an explanatory, and a descriptive one. But before I spell out that matrix in the next section, I add a bit more on the extension and relationship between humankind and humanity.

Humankind and Humanity Extending in Time and Space and Overlapping
Both groups—humankind and humanity—have vague boundaries and a historical dimension: they persist with change over time. As already noted, "humankind" refers to a biological group (H), that is, to being human in the sense of being a member of the species *H. sapiens*. "Humanity" refers to a social group (S), that is, to being human in the sense of being a member of a socially defined group. Nevertheless, the two groups overlap, as figure 1.1 illustrates.

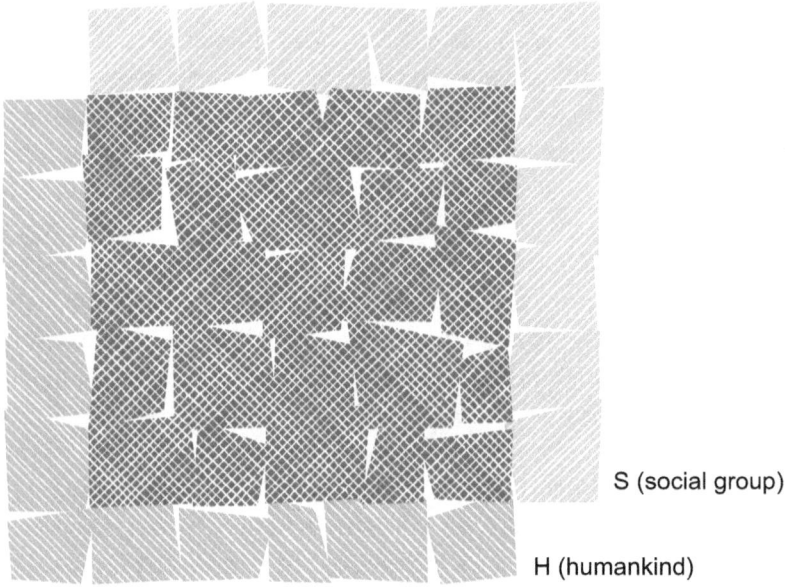

Figure 1.1
Humankind (H), a biological group, and humanity (S), a social group, as overlapping categories. The darker-striped group represents humankind (H) and the lighter-striped group humanity (S). The criteria for membership in these groups fall apart, but there is a great overlap of the resulting groups (the checkered and darkest area), since most individuals are members of both groups. (Illustration: Vera Brüggemann, Bielefeld.)

Having humankind and humanity distinguished in that sense not only helps to cast the net a bit wider than usual in the philosophy of science, but also to make sense of the intuition that humanoids might well be human in the social sense (persons) and get human rights, such as to vote, without being humans in the biological sense (*H. sapiens*). Recall movies dealing with humanoid creatures, for example, the movie *Bladerunner,* which seems to suggest that there is something unfair in excluding a humanoid creature from the moral group—those granted, for instance, bodily integrity—if the respective humanoid creature is cognitively, emotionally, and morally on a par with us and thus interacts with other humans in adequate and ethically and morally respectable ways—as the main female robot character in *Bladerunner* does.

In this book, I do not defend that humanity should include such humanoids or that all *H. sapiens* should be part of the moral circle, even though I believe that this is how it should be. Rather, I take it as more or less for granted. Toward the end, however, in part III, I add a few normative reflections on this issue, which runs under the label "moral standing." I do so also to prevent the impression that the account developed here assumes a kind of speciesism analogous to racism. Nevertheless, to normatively justify issues of moral standing, and thus who deserves certain rights, by far exceeds the scope of this study. It would require a detailed argumentation with respect to issues about human rights, killing at the margins of life (e.g., abortion and assisted suicide), and the ethics of care. Such a detailed account cannot be provided here, even though I clarify the connections to these issues in part III.

1.3 Three Different Concepts of Human Nature in Overview

Three Different Natures
The positive account that this book defends can be summarized in the following way. There are in the world three kinds of natures:

- *Classificatory nature*. With respect to humankind (i.e., the biological species *H. sapiens*), there is a classificatory nature, replacing a species' definitional essence: human nature consists of the conditions under which an organism partakes of the species *H. sapiens*. I will defend that these conditions are relational, referring to the genealogical nexus (and only derivatively to specificity or typicality of properties). In addition, there are membership conditions for humanity. Membership in what I call humanity does not depend on any biological connection; it depends on what those who are already part of it decide to regard as essential for their life form, which is then applied to others.

- *Descriptive nature.* Each of the groups not only has membership conditions (a classificatory nature), but embodies a descriptive human nature, the characteristic life form of the respective group, which I call *humanness*. Humanness uses the typicality aspect, but with a temporal dimension: it comprises traits that are typical for past and recent humans. I will defend that it refers to traits that are typical and stable—for humankind or for humanity, depending on case. Stability uses the aspect of fixity. Human nature in the descriptive sense thus refers to stable generalizations about a group of individuals (i.e., humans biologically or socially specified). Which generalizations are picked out also depends on which perspective (in particular, which science) is using them: an opposable thumb is important in anatomy, but less so for the psychologist, for example. In addition, only some sciences care for species specificity. Specificity is of special interest if the aim is to establish an animal-human divide. Philosophers, for instance, are often interested in defending a special status for humans in contrast to animals by pointing to species-specific properties.
- *Explanatory nature.* Humanness (in relation to humankind or humanity, depending on choice of group) needs to come from somewhere, that is, the respective descriptive nature needs to be explained by reference to ontogenetic development and evolution. The term *human nature* can thus also refer to explanatory factors—to an explanatory human nature. In such a case, the aspect of fixity becomes even more central: in explanatory contexts a rather fixed human nature has always been contrasted with a less fixed human nurture (i.e., culture, society, or environment). A core problem is that such fixity claims seem to conflict with the contemporary interactionist consensus that says that nature and nurture interact developmentally and evolutionarily. To solve that and related problems, the explanatory nature of biological species will be defended as being not internal to individual organisms but internal to the population of individuals making up the respective group (humankind or humanity).

Epistemic in Method and Post-Essentialist, Pluralist, and Interactive in Outcome

By distinguishing among these three different natures, I distinguish three different epistemic roles that the concept of human nature can play: a classificatory, a descriptive, and an explanatory epistemic role.[7] Considering these epistemic roles also allows showing that by classifying and explaining humans, we might even change them by way of so-called looping effects: the individuals classified or explained react to these epistemic acts with acts in real life and change themselves or other humans thereby. This is ultimately changing the generalizations that can be made with respect to the descriptive and explanatory nature.

Considering these epistemic roles separately also permits finding a way out of the traditional essentialism that comes with the concept of human nature. Essences, traditionally conceived, fulfill all three epistemic roles simultaneously: there is one thing in the world—the essence—that does it all. In the post-essentialist, pluralist, and interactive picture defended here, there are different things in the world, each fulfilling only one epistemic role.

In a nutshell, the account defended here is post-essentialist since it eliminates the concept of an essence. It is pluralist since it defends that there are in the world different things that correspond to three different post-essentialist concepts. There is not one thing in the world that corresponds to one concept of human nature; there are, rather, different things in the world (relations between individuals, humanness itself, and a set of especially important causal factors explaining humanness) corresponding to different post-essentialist concepts of human nature, used in different contexts, performing different epistemic roles. The account is interactive since nature and culture interact at the developmental, intergenerational, and evolutionary level and since humans change their human nature over time via classificatory and explanatory looping effects. The account developed on this basis is nonetheless a realist one since these natures exist out there in the world, mind-independently. The post-essentialist account is thus antimonist but not antirealist.

Since the term *human* can be understood to refer to either a biological group (a species) or a social group, the three natures appear twice over. That makes six useful ways of using the term *human nature*. The old essentialist human nature is dead; long live human natures in the post-essentialist plural!

Normativity Kept at Bay

There is also a normative pragmatic function of the concept of human nature, using the normalcy aspect. That aspect has regularly been used to divide people into groups in a normatively laden manner. Human nature is then taken to refer to those properties that humans should realize, for example, in order to survive or flourish, or even in order to be morally good. A nature has not only been taken to decide about moral group membership (who has duties or deserves rights in a certain sense); the normative nature also specifies which rights are included, based on what is normal for being human. If such a normative nature exists, then there is a natural goodness. It is evident that this mainly uses the fourth aspect mentioned, normalcy, connected to typicality, maybe even fixity. It has a dark side, since it ultimately leads to dehumanization, as chapter 2 illustrates in detail.

No normative aspect of this sort will be part of the post-essentialist, pluralist, and interactive account. Nevertheless, I say something about it in part III, where I reduce the alleged normative nature to the result of contestations that we have about our nature. It is dismissed as a mirage of us contesting how the classificatory and descriptive nature should develop over time, which involves not only projecting the latter into the future but also classificatory and explanatory looping effects—effects in reality that result from reflexivity.

The goal in this procedure is to keep the normativity attached to discussions about human nature at bay. This is not in order to ignore it, but in order to see better where it legitimately has something to say. I more or less assume the following: it has nothing to say with respect to humankind and the scientific image that can be defended about humankind's nature in the classificatory, descriptive, and explanatory sense. This is why it will not be of concern in part II, which focuses on humankind and the concepts of human natures attached to it. I justify this assumption in the form of a dispensability proof: no normativity is needed to fulfill all three epistemic roles that the traditional concept has played in sciences, given contemporary knowledge about evolution, heredity, and development. If so, then parsimony (as an epistemic value) tells us to leave it out.

Introduction

Let me sum up this brief outlook. After discussing three challenges in part I, I develop a positive picture in parts II and III. The result will be that there are in the world two times three natures: two times (for humankind as well as humanity), there is a classificatory, a descriptive, and an explanatory nature. Although connected by overlap of groups, conceptual aspects used, and relations between, these six different concepts of human nature do not refer to the same facts in the world. They have different extensions and use a different set of the four core aspects mentioned. Nevertheless, they all exist. Details on these natures will have to wait for individual chapters of part II, and normative issues will be discussed in part III. Part I analyzes why the concept of human nature has such a bad reputation in contemporary philosophy.

1 Three Challenges

There is a vernacular and a scientific concept of human nature. The vernacular concept of human nature is used (among other things) for dehumanization, that is, to conceptually include and exclude people as more or less human. Given that different people fill the concept with different content, the vernacular concept is socially perspectival: depending on social position, something else is regarded as part of human nature. In science, by contrast, human nature is used in a more objective sense. In addition, talk of a species' nature (be it human nature or any other species' nature) has often been taken to refer to an essence that is simultaneously classificatory, explanatory, and descriptive for the respective kind. Finally, talk about nature traditionally involves a contrast to nurture (or culture).

Dehumanization, essentialism, and the nature-nurture divide are three major reasons why the concept of human nature was under attack in the twentieth century. Chapter 2 deals with the dehumanization challenge: that the concept of human nature is used socially perspectivally, to humanize those speaking and to dehumanize others. Chapter 3 deals with the Darwinian challenge: that the contemporary Darwinian ontology challenges philosophical assumptions about natures as essences. Chapter 4 addresses the developmentalist challenge: that nature cannot be separated from nurture. By introducing these three challenges in detail, part I prepares the ground for the more constructive part II.

2 The Dehumanization Challenge

This chapter analyzes the vernacular concept of human nature: how the concept of human nature is used outside of science.[1] It shows that in social affairs, it is used for dehumanization—for regarding some people as not or as less human.[2] It argues that the vernacular concept of human nature is socially perspectival: what is part of human nature (and thus fully human) depends on who speaks. The concept of human nature (taken as a descriptive one) is filled—depending on context—with different content (i.e., different ideas about what it means to be human), but always with what those speaking consider as their essential characteristics lacking in others.

For a philosophical understanding, it follows that a general account of the vernacular concept of being human and (derived from that) of human nature needs to be functional: focusing on the pragmatic function of the concept—that it is used for dehumanization, and that means for regulating social inclusion and exclusion. Focusing on any fixed content would miss the point of the vernacular concept. The challenge for a balanced philosophical account of human nature is then twofold: there is an epistemic challenge to overcome the social perspectivity in order to arrive at an objective account of human nature, and there is a moral challenge to keep dehumanization in check.

The claim that the concepts of being human/human nature are used for social demarcation and dehumanization is not new. It can be found in historical and philosophical literature (e.g., Smith 2011; Jahoda 1998; Clark, Golinski, and Schaffer 1999; Mikkola 2016) and in the empirically working contemporary social psychology (e.g., Haslam and Loughnan 2014). Since there is not sufficient integration of the literature and thus no clear sense of what follows philosophically for a general account of the concept of human nature between science and society, a systematic account of dehumanization and the respective politics of human nature needs to be developed.³ Parallels to the concepts of race and gender exist, since these concepts also facilitate social inclusion and exclusion. For lack of space, however, the comparison of similarities and differences between these sortals cannot be analyzed here.

I begin in section 2.1 by introducing how normalcy assumptions and dehumanization enter the vernacular concept of human nature. In section 2.2, I distinguish general features of dehumanization and two forms of it. Together with the examples mentioned, this allows me to specify in section 2.3 the main thesis of this chapter: that the vernacular concept of being human/human nature should be understood functionally because of its social perspectivity. I close in section 2.4 by describing in more detail the epistemological and moral challenge that derives from it for a general account of human nature that crosses the boundary between science and society.

2.1 The Vernacular Concept of Human Nature

An important pragmatic political function of the concept of human nature is to facilitate social inclusion and exclusion: it is involved in regulating who is us and who is them, a process by which some people get dehumanized. As Hull (1986, 7) stressed in his famous critique of the concept of human nature, "The normal state for human beings is to be white, male heterosexuals. All others do not participate fully in human nature."

To participate fully in human nature, an individual has to be *normal*. That the normalcy aspect is fulfilling a political function (not an epistemic role) is suggested by the fact that we mainly use it for our species. In the words of Proctor (2003, 220), one does not ask about an entity "being 'fully cockroach' or 'fully chimpanzee.'" Hull (1986, 6) also stressed this long ago: we describe other species often in careful statistical and descriptive manners, but when it comes to our species, we often fall back into essentialist traps, involving normalcy and that entails normativity. Hull regarded this "coincidence [as] highly suspicious."⁴ The statement "humans are heterosexual" is a normatively laden normalcy claim (in the linguistic form of a generic) with essentialist, variation-discounting overtones. It has caused considerable harm, psychological and physical, for those not conforming. By contrast, we do not care about chimps fully or less fully realizing chimpanzee nature; presumably this is because we do not have (at least not yet) relevant social relations with chimps, within which the same social, political function could kick in.[5] At least so far, we are appealing to a nature in a normatively laden and less recognition-causing manner only in the case of other humans. Since epistemically viewed, studying different species is the same business, the source of the exceptionalist way of dealing with our nature in a normative manner lies first and foremost in its political function, not in its epistemic roles.

A concept of human nature that facilitates dehumanization, a phenomenon that I analyze in detail below, also shows up in a variety of philosophical theories that take human nature as a source for normativity. There is the tradition of natural law theories, which ground positive law in a natural law.[6] There are theories of natural goodness and human flourishing that use human nature to christen certain traits (and certain behaviors) as "good" because they are "natural."[7] With respect to enhancement, the normativity attached to the concept of human nature is used to distinguish between legitimate and illegitimate enhancements—interventions meant to improve body or mind. Theories of human rights might also be conceptualized to rest on a normative conception of human nature. This list could be completed, systematized in different ways, and there are different directions from which critique against these traditions can be raised. To directly address them would by far exceed the possibilities of this book.[8] The important point for this book is something more basic: one (not necessarily the only) theory-independent source of all these versions of employing a normative concept of human nature is that certain properties or behaviors are discriminated against and excluded as not good. Together with this, certain kinds of people, those exhibiting these properties or behaviors, either are excluded from a certain social status or are denied certain rights. Nussbaum (1992, 209) called it the problem of "prejudicial application" of human nature.

These normative uses of the concept of human nature in philosophy once again show that human nature is connected to social demarcation and exclusion, which is at the heart of dehumanization. People can be excluded from certain status or rights by regarding them or their behavior as not human, less human, or unnatural.[9] On the basis of dehumanization, some people are then depicted as brutes, vermin, demons, lice or as mere objects of desire, and so forth. Ultimately dehumanization involves prejudice and stigma and can lead to abhorrent, if not deadly, consequences for those dehumanized.

2.2 Dehumanization Systematically Viewed

Dehumanization cannot happen without a concept of human nature in the minds of those dehumanizing others. In other words, dehumanization has a cognitive, conceptual dimension: what makes us human (i.e., what it means to be human for us) is what makes others less human. To regard others as less human, one needs to assume (implicitly or explicitly) what it means to be fully human. For this reason, there is a dependence of dehumanization on a concept of being human and with that on a concept of human nature in the minimal sense of denoting important typical features of the human life form, typical traits of humans that some lack. Dehumanization can then be defined as an evaluative stance (merely cognitive or also behavioral) toward other humans that consists in drawing the line between individuals or groups (as in-group/out-group) according to an assumed concept of what it means to be human.

Thus, even if all forms of dehumanization that target a group of people are cases of in-group favoritism, not all forms of in-group favoritism are cases of dehumanization. Fast-changing peer group fashion codes might be an example of in-group favoritism without a connection to the concept of human nature. A process of inclusion and exclusion must have something to do with the concept of being human to count as a case of dehumanization. Otherwise the concept would become too broad.[10]

The core of dehumanization is a belief (seemingly factual, such as a projection or an overgeneralization, as in standard stereotyping) with respect to being human: a belief that Jews are evil people (whereas normal humans are not); that women are childlike (whereas the normal grown-up human is not); that black men are aggressive rather than civilized (in contrast to their white counterparts). These seemingly factual beliefs already contain an evaluation, but the evaluation can also be an extra step. The evaluation is what leads to prejudice—hate, pity, disdain, superiority, fear, anger, or envy, for example.

A belief (with an evaluation) is a cognitive stance that can (but does not have to) lead to behavioral consequences. Dehumanization thus often leads to diverse kinds of discrimination (social exclusions of ethnic groups, women, or homeless people, for example) and violence (such as rape, war, or genocide).[11] Because of that, socially important issues are connected to dehumanization, such as racism, sexism, eugenics, disability, and poverty, as well as violence-related issues such as genocide and other forms of atrocities related to group conflicts. Haslam (2006) writes, "At the milder end of the spectrum, such [behavioral] consequences will not involve hatred and aggression, but may include patronizing and condescending behavior, dismissive attitudes, lack of empathy, and indifference to the interests of others." At the extreme end of the spectrum can be outright exclusion from the moral realm and, as a consequence, violence. Dehumanization then involves regarding people as "outside the boundary in which moral values, rules, and considerations of fairness apply," legitimizing murder, rape, torture, and other atrocities (Opotow 1990, 1).

To sum up so far, what we think about being human and thus human nature influences how we treat other beings. We do not only employ the concept of human nature for us; we regularly use it to deny humanness to them, that is, other individuals or out-groups or subgroups. Standard examples for such targets of dehumanization are women, behaviorally deviant people, disabled people, refugees, strangers, pariahs, other ethnic groups, and enemies.

Historians, cultural anthropologists, and social psychologists have accumulated plenty of evidence for this dark side of the vernacular concept of human nature. To illustrate the dependence of dehumanization on the concept of human nature and prepare the ground for the thesis about the social perspectivity of the vernacular concept of human nature, I discuss two examples in detail in this chapter: the dehumanization of women and the dehumanization of non-Europeans. This will also allow me to distinguish between two forms of dehumanization.

The Dehumanization of Women

The dehumanization of women has a long history. Aristotelian essentialism, for instance, implied that variations in a species (e.g., *H. sapiens*) are deviations from a type. Human nature in Aristotle can be understood as referring to the human life form, which is not only the form (contrasted with matter) but also the end (*telos*) of human flourishing. Deviations are conceived as not fully realizing the form of the type. They occur because something interfered. Furthermore, deviant individuals are regarded as inferior to those more closely realizing the form. Form is norm. Women were for Aristotle (infamously in *Politics*, Book 1, 1252a-1260b) such inferior deviations—deviations from human nature and inferior to the free men representing the type.

To think of certain kinds of people as further away from a specific life form of a species involves, at least in the Aristotelian picture, a distinction between proper and improper kinds of developmental causes. If a female is produced during biological reproduction, as Aristotle writes in the *Generatione Animalium* (767b7–25), then "nature has in a way departed from the type." He acknowledges that for sexually reproducing species, it is a "natural necessity" that females are regularly produced; nevertheless, when they are produced, the embryo is "deficient," since if "the generative secretion in the catamenia is properly concocted, the movement imparted by the male will make the form of the embryo in the likeness of itself."[12] The form of the embryo in Aristotle's picture is due to the male semen alone; women contribute matter only, so they cannot make an embryo in the likeness of themselves since, lacking form, they are not like anything and contribute matter only. Without interfering causes, there would be no variation and only well-functioning Greek men, realizing in full grandeur the nature of being human, representing the natural state of being human, that is, how one *should* be.

This explanatory schema is teleological and essentialist. Roughley (2011, 13) describes the teleological aspect: "Nature in this sense has dispositional to-be-realizedness, i.e., in the absence of defeating conditions, the natural entity will realize its full and specific form." The schema is essentialist since some traits (such as rationality) are picked out as essential, contrasted with others that were deemed to be negligible for the essence of being human.

As Deslauriers (1998) and Witt (2005) stress, Aristotle still regarded women (despite them being deviants) as (what we now call) members of *H. sapiens*, that is, as partaking in the same species as Greek men.[13] It seems that variation within humankind did not prevent Aristotle from including deviants as members of the one humankind and as consequently partaking in the same human nature. The deviants were *same same but different* (as an Asian idiom goes).

In terms of dehumanization, one can capture the case as follows: it is a case of attributing more or less of humanness to a particular group in an overgeneralized manner, which very likely involved some emotional evaluation and definitely had some behavioral consequences. After all, in ancient Greek society, men were supposed to be the masters of women and women had quite restricted rights.

By the nineteenth century, not much had changed—except two things: the interpretation of the properties women were believed to lack and the essentialist metaphysics beneath. Herbert Spencer, for instance, founder figure of sociology, far from Aristotle's metaphysics, wrote about the smaller body height and smaller brains of women. Their smaller brain is presented as a natural necessity without which women would be unable to give birth. Because of their smaller brain there is a "perceptible falling-short" of women in the "two faculties, intellectual and emotional, which are the latest products of human evolution—the power of abstract reasoning and that most abstract of the emotions, the sentiment of justice—the sentiment which regulates conduct irrespective of personal attachments and the likes or dislikes felt for individuals" (Spencer 1873, 374).[14] Even if Spencer's concrete form of dehumanization of women is not likely to arise in contemporary sciences, women are still regularly dehumanized, even if often only implicitly and not with respect to the same properties: "abstract reasoning" and the "sentiment of justice." The content—the essential properties picked out—changed, but the function of dehumanization remains.[15]

The Dehumanization of Non-Europeans

The explicit dehumanization of Non-Europeans, e.g. Africans and Amerindians, is another well-known form of dehumanization, important in the context of colonialism and prevalent well into (if not way beyond) the nineteenth century. After Columbus's landfall in 1492, Spaniards hacked off natives' limbs, burned them alive, and fed babies to the Spaniards' dogs. Some complained, including the Jesuit priest Montesinos, who asked "Are these not men? Have they not rational souls?" Even in the sixteenth century, humanist scholars such as Giordano Bruno and the alchemist Paracelsus, among others, denied a shared ancestry of *their* people with *those* Amerindians, by regarding them as non-Adamic in origin (as not descending from Adam, the common denominator of the Judeo-Christian-Islamic androcentrism), or as homunculi (beings with a human body but no soul).[16]

In the nineteenth century, when Charles Darwin was traveling on the *Beagle*, it was still a commonsense position in Jamaica or Brazil to regard non-European people as a separate species, an intermediate step between apes and humans, as Desmond and Moore (2009) report in their biography on Darwin. Darwin, a member of a family fighting for abolition of slavery, had thus a "sacred cause" to write his *On the Origin of Species*: to develop a theory of evolution that relied on common descent, to show that the racism underpinning slavery is scientifically wrong. Darwin's "*human* project" was "foundational," Desmond and Moore (2009, xvii) claim:

> While activists proclaimed a "crusade" (his word) against slavery, he subverted it with his science. Where slave-masters bestialized blacks, Darwin's starting point was the abolitionist belief in blood kinship, a "common descent." Adamic unity and the brotherhood of man were axiomatic in the anti-slavery tracts that he and his family devoured and distributed. It implied a single origin for black and white, a shared ancestry. And this was the unique feature of Darwin's peculiar brand of evolution.

Desmond and Moore (2009, xix) regard it as historically "paradoxical" that Darwin's theories "have been used to justify racial conflict and ethnic cleaning," given this background. From the point of view of dehumanization studies, it is evidence that dehumanization is so persistent that its effect is not eradicated by Darwin's motivation and account. Having every human included into one species still allows to regard some of them as less human, as less evolved.[17]

Contemporary experimental research shows how persistent it is. Kteily et al. (2015) show that out-groups are regularly and explicitly depicted as less evolved. Goff et al. (2008) show that there is also an implicit and direct association between black people and apes. Their first experiment, a degraded object test, produced evidence for an implicit bias that they name "black-ape facilitation effect." Study participants were first subliminally primed with black, white, or neutral faces. They then had to recognize degraded line drawings of animals that became step by step easier to identify. The prediction was (for white and nonwhite study participants) that "exposure to the Black male faces would facilitate identification of the ape images, whereas exposure to the White male faces would not." Debriefing confirmed that participants were not aware of priming faces. Completion of the Modern Racism Scale (MRS) and the Motivation to Control Prejudice Scale (MCP scale) were used to make sure that the results are independent of individual differences in explicit antiblack prejudice or attitudes about prejudice. One result is illustrated in figure 2.1.

Further studies added the following points: first, the effect is bidirectional (images of apes also direct attention to black faces) and, second, the effect is specific for black people rather than a general out-group bias, since analogous tests with Asian faces did not show the same results. By using colorless line drawings, the researchers tried to rule out that the effect is simple perceptual color matching. By also testing associations with other animals (e.g., big cats such as lions that are also stereotypically perceived as aggressive), they tried to rule out that what they observe is simply an association between black people and aggression rather than a specific association between black people and apes.

Nevertheless, aggression seems to play a role in the association, as two final studies suggest. These showed that not only perception and cognition are influenced by the black-ape association, but also judgment about violence against black individuals (compared with violence against white individuals). First, they studied whether priming with ape images or big cat images influences the judgment of whether police violence against suspects (shown to study participants in a video) is justified. Among the results was that violence against black people was more likely to be regarded as justified when primed with ape images. Their final historical (rather than experimental) study on death sentences in Philadelphia between 1979 and 1999 adds evidence in the same direction. Goff et al. (2008, 304) conclude from this last study that "Black defendants are more likely to be portrayed as apelike in news coverage than White defendants and that this portrayal is associated with a higher probability of state-sponsored executions."[18]

Explicit, Implicit, Graded, and Two Forms of Dehumanization
In the twenty-first century, it is (I hope) unlikely that somebody explicitly and literally regards women or non-Europeans (or anybody else) as (1) members of a different species or as (2) like apes.[19] Thus, explicitly and literally (1) denying membership in humankind and (2) completely denying humanness should be rather rare. An exception (even today) is probably war propaganda and caricature, as in the famous war poster in figure 2.2, which used a quite old "nexus of sexism and racism" (Hund, Mills, and Sebastiani 2016).

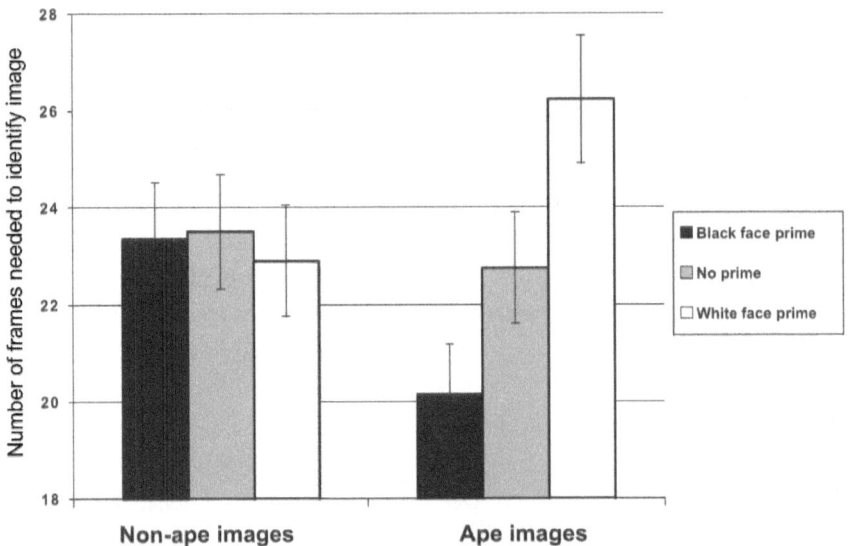

Figure 2.1
The black-ape-facilitation effect, according to Goff et al. (2008). Participants had to identify images (non-ape images or ape images) that become easier to identify from frame to frame (vertical axis). When primed with black faces (black columns), significantly fewer frames are needed to identify ape images (right side), compared to the case when non-ape images had to be identified (left side), and compared to when study participants are not primed (gray columns). Seeing a black face facilitated ape-image recognition. When primed with white faces (white columns), significantly more frames are needed for viewers to recognize the ape images (right side), compared to the non-ape images (left side) and compared to when not primed (gray column). The conclusion is that "mean frame number at which the animal could be detected" is "a function of animal type and race prime," which is taken as part of what Goff et al. call the black-ape-facilitation effect. (Reprinted from Phillip Atiba Goff, Jennifer L. Eberhardt, Melissa J. Williams, and Matthew Christian Jackson, "Not Yet Human: Implicit Knowledge, Historical Dehumanization, and Contemporary Consequences," *Journal of Personality and Social Psychology* 94, no. 2 [2008]: 292–306, 296. With permission from publisher APA.)

Figure 2.2
The rapist enemy. This propaganda poster (1917–1918) combines the dehumanization of women (as weak objects of desire) and an enemy (as the aggressive animal-like rapist). The enemy in this case was World War I Germany. (Artist: Harry R. Hopps, Lithograph Print, Collection of the Library of Congress. Public domain.)

In addition to an ongoing explicit dehumanization, as evidenced in Kteily et al. (2015) and in propaganda and caricature, people are implicitly regarded in all kinds of contexts as less human, and in the two forms just mentioned— as (1) less human in heritage (relational dehumanization) or (2) with respect to humanness (i.e., with respect to properties such as rationality or civility versus aggression or any other putative essential property of being human). Membership in the humankind or instantiation of so-called essential properties is then not denied but graded.

1. *Relational dehumanization.* This kind of dehumanization refers to genealogy or social interaction (and only subsequently to qualitative properties). What counts is where individuals come from or where they socially belong to. This relational dehumanization demarcates the boundaries among people along the lines that relations between people allow to be drawn (rather than directly along the lines of similarities of properties). Thus, some groups of people were regarded as not belonging to humankind (or humanity) because they were regarded as non-Adamites. Also, when some people trace their genealogy and take it to be important for their identity, it can well happen that no similarity will make up for the closeness or distance a certain individual has genealogically to the ones believing in genealogy being important. Genealogy trumps similarity in such cases. It is important to note that an individual can be more or less related to other individuals (depending on how deep in the past one finds the last common ancestor of the respective individuals), even if there are no races in the sense of genealogically separate groups (e.g., since there is enough interbreeding or even panmixia) between populations so that the groups are not genealogically isolated. Thus there can be relational dehumanization, without racism in the narrow sense of group-based discrimination.

2. *Property-based dehumanization.* Talk about some humans being "normal" (while others are abnormal) often happens with reference to qualitative traits such as rationality or morality. We use a concept of humanness, a list of properties deemed to be unique or typical of those identified as humans (on social or biological grounds, depending on case) and allocate it differentially to different (groups of) people. Thus, property-based dehumanization can be characterized as a content-based discrimination against people, whereby the content is generated by reference to the content of a concept of human nature (in the

descriptive sense). Aristotle's dehumanization of women and slaves as less rational is a case in point for that kind of dehumanization. More or less close genealogy was completely irrelevant for that property-based dehumanization. In an Aristotelian picture, a daughter is less human than a son, despite equal genealogical relatedness to their parents.

Relational dehumanization is likely to be less important in the West today given that humans interact biologically and socially on a global level (actually, they have for quite some time) and given that a universalist concept of humankind (there is only one species of humans, all are included) has taken hold and has done so very likely also because of the global interactions. Yet historically viewed, relational dehumanization was an important issue and, according to an important school in cultural anthropology, relationalism, it is still an important scheme of thought in alternative ontologies in groups of people not yet Westernized (see Descola 2005; Viveiros de Castro 1992, 1998; Lloyd 2011, 2012). In some of these alternative ontologies, it holds that if somebody does not belong to a group (i.e., relates or interacts with other members of the group in inadequate manners), that person is regarded as not human (or less human, if interaction comes in degrees), independent of intrinsic properties.

Actual cases of dehumanization practices might well involve both forms, relational and property-based dehumanization, and it might be hard to decide which form is active (or more active), not only but also because one kind of dehumanization can be a proxy for the other. For instance, the results from Goff et al. (2008), especially of the last two studies, can be explained either by a relational exclusion attitude of *black-means-apelike-means-less-human-by-heritage* or by a specific triadic association of *black-means-apelike-means-aggressive-and-therefore-exibiting-less-humanness*. It is unclear from the data whether the dehumanization of black people happens because there is an implicit assumption that black people—by heritage—are more closely related to apes and therefore less evolved than the rest of humankind, or because there is an implicit assumption that black people are more like apes. In other words, it is unclear whether genealogy or similarity is driving the association and thus forming the rationale behind the black-ape association.

Similarly, when early hominids, devoid of complex tools, taken as evidence for lack in creativity, are regarded as less human, this dehumanization can happen because of the weakness of relation between contemporary humans and these early humans (the trait "complex tool use/creativity" would then merely serve as proxy, i.e., as indicator of strength of relation), or it might be just the other way around: that early hominids are regarded as less related because they do not have the same properties as we. Proctor (2003, 220), from whom I take the example, makes a specific suggestion toward the second interpretation: that we regard them as not fully human because they were not as innovative as us. He even adds that this might stem from a capitalist prejudice toward novelty: "Incessant innovation is not an obvious prerequisite for being human: this may be a modernist prejudice, if not a capitalist imperative."

The barbarians of the Greeks were considered enemies or at least outsiders (without sufficient positive social relations with the Greeks), and at the same time they were regarded as without civilization, literally those without proper language, being able to utter "ba-ba" only, whence the label "barbarians." What was driving the dehumanization of the barbarians: the relations or the properties? Hard to tell.

In sum, it might often be hard to decide which form of dehumanization is taking place, but for this study, this is not decisive. What is important is only that they can fall apart and that both can happen. I take it that implicit (if not explicit) dehumanization in both forms, relational and property based, is still among us and likely to be very widespread. I take the evidence accumulated in history, anthropology, and social psychology, reviewed in Haslam and Loughnan (2014), to confirm this.[20]

2.3 Social Perspectivity

What is the upshot of dehumanization for understanding the concept of human nature? I started with the claim that each case of dehumanization involves assumptions about what it means to be human—that the respective concept of human nature is filled with different content that is then unevenly attributed to different people, with the speaker usually in the center as fully human.

The important point is that the content is completely exchangeable and perspectival. The examples I discussed confirm this: the content of the concept of human nature changes historically and varies between people and scientific contexts, yet the function of the concept of human (and, with it, human nature) endures: it is used for dehumanization. In particular, the examples I have used above illustrate:

- A change from Aristotelian metaphysics to the nineteenth-century naturalist philosophy of Herbert Spencer with respect to the dehumanization of women
- A change from colonial dehumanization of Amerindians as non-Adamic in heritage or as homunculi, to the black-ape-facilitation effect
- That different traits can be used to liken somebody to apes (compare Spencer's two traits with the aggressiveness attribution involved in the black-ape-facilitation effect)

Further examples would simply add further variation in content, that means properties discounted from some people. Even the underlying ontology of how to distinguish between humans and animals, and also which form of dehumanization is involved, varies. But despite changes in content and form of dehumanization, the function stays: social inclusion or exclusion, including this kind of people and excluding that kind of people, regulating who is us and who is them.

I conclude from this that the content of the vernacular concept of human nature as such, as well as the specific form of dehumanization, are historically changing, but the pragmatic function, dehumanization, is constant. Thus, if one aims to understand the vernacular concept, a functional perspective is necessary.[21]

Converging Evidence

I take the results of experimental studies on dehumanization in social psychology to converge on such a functional perspective, despite differences in details, as described in overview by Haslam and Loughnan (2014). I also take the history of theorizing about dehumanization to confirm that perspective. Hume, in his *Treatise of Human Nature*, as part of his analysis of love and hate (book II, part 2, section 3), discussed sympathy as a bias that favors members of one's own group and dehumanizes others. The anthropologist and sociologist Graham Sumner (1906, 12–15) then discussed dehumanization under the label "ethnocentrism" (his technical term for in-group favoritism) and stressed how widespread it is, since "as a rule, it is found that native peoples call themselves 'men.' Others are something else perhaps not defined—but not real men." Sumner mentioned a few examples for (1) the relational form of dehumanization. He also gave examples of groups that exhibit a less strong dehumanizing ethnocentrism, by believing in various kinds of superiority with respect to (2) property-based dehumanization. Among the examples are the Chinese, Greeks, Romans, and Arabs of antiquity. Lloyd (2011, 834) writes that in antiquity, "Chinese names for many non-Han tribes incorporate the radicals for animals, especially the dog." These groups of people were believed either to be descendants of outsiders (rebels or mavericks that got ostracized) or to stem from a liaison of a Chinese princess with a dog, as Stichweh (2010, 25) reports. Lévi-Strauss (1952, 12) famously put the mindset of ethnocentrism as follows: "Humanity is confined to the borders of the tribe, the linguistic group, or even, in some instances, to the village." Sahlins (2008) mentions that the Chewong hunter-gatherers of Malaysia include certain nonhuman animals as humans and regard certain *H. sapiens* as animals. Who is human depends, in their "alternative ontology" (Descola 2005), on relations: whom one knows and on whether there is social interaction or social dependence involved. Ingold (2000) also reported cases of such relational ontologies.

Reciprocity

Finally, Lévi-Strauss stressed an important reciprocity in dehumanization attitudes:

In the Greater Antilles, some years after the discovery of America, while the Spaniards sent out investigating commissions to ascertain whether or not the natives had a soul, the latter were engaged in the drowning of white prisoners in order to verify, through prolonged watching, whether or not their corpses were subject to putrefaction. (Lévi-Strauss 1952, 12)

While the Spaniards were skeptical about the humanness of the natives in terms of "having a soul," the natives, it seems, were skeptical about the humanness of the Spaniards in terms of "having life," tested by whether the decaying pattern characteristic for life occurs. Haslam (2006) would call the one *animalistic dehumanization* and the other *mechanistic dehumanization*, two forms of dehumanization that occur, according to him, repeatedly and in various contexts, depending on which properties are at issue.

Compare this with a report on how those suffering from racism can reciprocate with dehumanizing the racists:

Lucky for us we lived very far in the country. We saw very few white people. And when we went to town we followed rules about where we could go. And we just followed our parents. They basically helped us to see white people as, you know, very stunted. That was just the way they were. There was nothing you could do about it, they were just like that. (Who knew why they were like that?) And that was helpful. *They were discussed as if they were the weather.* (Alice Walker, quoted in Fricker 2016, 169, emphasis added)

Regarding someone as "like the weather" is a case of mechanistic dehumanization: the other is thought of and treated like an object rather than like a subject.

Reciprocity shows that what is accomplished by dehumanization is creating distance, then boundaries, and ultimately exclusion. Again, only by assuming a radical social perspectivity in the concept of what it means to be human can one explain the reciprocity that is exemplified in these examples.[22]

Social Perspectivity
The historian Koselleck ([1993] 2006, 279) similarly claimed that the concept of being human is, historically viewed, a "blank mold" (*Blindformel*). According to Koselleck, asymmetric concepts for collectives, such as the Christians, carry with them a contrast – a concept for enemies (*Feindbegriffe*), for example, the non-Christians. The content of such an asymmetric and contrastive concept for collectives (what it means to be a non/Christian) is, however, not as exchangeable as the content of the concept of being human. The concept of the human is a conceptual blank mold since any group can fill it with what that group is like and exclude the respective others as less human. Sahlins (2008, 2) points basically at the same idea when he summarizes the dehumanizing mindset as a *l'espèce c'est moi* attitude: *I am the species; all others are less human*. Similarly, Louise Antony (2000, 34) writes that the word *humanity* in its vernacular meaning is an "empty honorific, used to indicate especially important or valuable aspects of our lives as we see it." I also take Smith (2013) to agree on this when he claims that *human* is a term that is indexically used, like the term *here*.

The core thesis of this section is thus that the content of what it means to be human varies with the perspective of those who speak. The vernacular concept of being human (and consequently the concept of human nature resulting from it) is characterized by a social perspectivity that is part and parcel of dehumanization.

2.4 The Challenge That Derives from Dehumanization

Dehumanization shows that the social meaning of the term *human*, as part of which it first and foremost refers to us rather than them, only contingently refers to *H. sapiens*. It only now refers to *H. sapiens* because those dominating the discourse (at least in the West) have accepted themselves as *H. sapiens*. Whether the group referred to is humankind or humanity is thus less important than understanding the function of dehumanization itself. The concept of being human/human nature can be used for inclusion or exclusion in groups and for respective normative judgments about being a good person/good human.

From a political and moral point of view, the challenge for an adequate account of human nature is to keep dehumanization in check to minimize its role in sciences and society. This dimension of the dehumanization challenge arises since dehumanization is usually evaluated negatively, as something that should be prevented. It conflicts with fundamental ideas about equality, justice, and human rights. If dehumanization should go, the concepts that further it should go too in order to prevent it. If the concept of a human nature furthers dehumanization, we should eliminate it. This eliminative conclusion is part, for instance, of Hull's (1986) famous critique of the essentialist concept of human nature.

The second dimension of the dehumanization challenge is epistemic: How can there be an objective account of human nature if there is all this social perspectivity involved? A highly socially perspectival concept cannot be the foundation of an objective, scientific account. A scientific concept thus needs to be disconnected from the vernacular concept. From the scientific point of view, the social perspectivity in the vernacular concept needs to be overcome in order to arrive at an objective concept of being human/human nature.

Can sciences provide objective content for a concept of human nature? They can, as I shall argue in part II, but in a pluralist manner only (using different concepts of human nature) and in a manner that fails to get completely rid of dehumanization, as I will argue in part III. Science cannot resolve the politics of human nature described in this chapter, but it can resolve the social perspectivity to a great extent at least. The next two chapters and part II are concerned with what science can provide for an objective concept of human nature.

3 The Darwinian Challenge

This chapter presents a debate about human nature as referring to the essence of the species *Homo sapiens*. The debate has unfolded since the 1980s as part of philosophy of biology, a branch of philosophy of science that has species (and thus species' essences) as one if its fundamental concepts. The debate shows that contemporary biological knowledge—related to evolution, development, and heredity—sets considerable constraints for a concept of human nature that refers to species as an evolutionary category.[1] I defend the claim that natures as essences are not needed, and in certain respects, they are even incompatible with contemporary Darwinian biology. With this, I reconstruct the core of an anti-essentialist consensus in philosophy of science that results from populational thinking, which became dominant in biology via Darwinian theory and statistical ways of understanding hereditary variation in Galton.[2] The anti-essentialist consensus will be specified with respect to two central epistemic roles of essences: their classificatory and their explanatory role.

I introduce in section 3.1 what essences would require if they existed for biological species. For this, I outline the concepts of essence, natural kinds, and species. I then describe the Darwinian challenge concerning the two core epistemic roles of essences, analyzing in detail the challenge for the classificatory role in section 3.2 and the challenge for the explanatory role of essences in section 3.3. Finally, in section 3.4, I briefly situate the anti-essentialist consensus in a broader picture about metaphysical essentialism and psychological essentialism, as well as broader debates about human nature.

3.1 What Essences Would Require

Essentialism, Natural Kinds, and Species

Sober (1980, 354) characterizes species essentialism with respect to human nature as follows: "The essentialist hypothesizes that there exists some characteristic unique to and shared by all members of *Homo sapiens* which explains why they are the way they are." Ereshefsky (2010a) splits up the second, the explanatory aspect, into an explanatory and a descriptive-predictive aspect:

> One tenet is that *all and only* the members of a kind have a common essence. A second tenet is that the essence of a kind is *responsible* for the traits typically associated with the members of that kind. For example, gold's atomic structure is responsible for gold's disposition to melt at certain temperatures. Third, knowing a kind's essence helps us *explain and predict* those properties typically associated with a kind. (Ereshefsky 2010a, emphasis added)

A number of authors, primarily Mayr (1963), Hull (1978, 1986), and Sober (1980), influentially debunked such an essentialism for the case of biological species. As a result of the anti-essentialism about biological species, talk about human nature as an essence in the above tripartite sense has had a rather bad reputation within philosophy of science. That human nature in the essentialist sense is dead is the message of the Darwinian challenge. I reconstruct the respective arguments in a systematic manner in this and the next two sections.

If one takes the term *natural kind* to refer to those kinds for which there is an epistemically tripartite essence, then the question is whether human nature, referring to *H. sapiens*, is a natural kind term. Are there (first tenet) necessary and sufficient conditions for partaking in the kind, that (second tenet) are at the same time explanatory for the way these entities are, and (third tenet) thus help in explaining, describing, and predicting other properties of the kind? According to the anti-essentialists, biological species are no such natural kinds since there are no essences of biological species.

Whether they are right or wrong certainly depends on how the concepts of an essence, a natural kind, and a species are spelled out. Despite consensus on the above characterization of essences and natural kinds as a foil, essences can mean quite different things, as Wilkins (2013) shows. The same holds for the concepts of species and natural kinds. Mayden (1997) lists twenty-two species concepts that are conventionally put into four groups: phenetic, biological, ecological, and phylogenetic species concepts.[3] Hacking (1991, 2007b) doubts that there is a coherent concept of natural kind shared across different debates about these kinds.[4]

Furthermore, some of the anti-essentialists (e.g., Hull 1978) would add that biological species are actually individuals (particulars) and not kinds. Whether that is so is of highest importance for understanding fundamental philosophical categories (such as "individual" or "kind") as well as fundamental biological categories (such as "species"). But for the purposes of this study, this issue can be put aside. First, it can be solved in a pluralist epistemic manner, as Rieppel (2013, 167) suggested: if one looks from the organisms partaking in the species, the latter appears as marking the extension of a kind, and if one looks from the species, the organisms appear as parts of a larger individual. Nothing empirical hinges on which way one looks at it.[5] In particular, nothing in understanding human nature depends on it, since what counts for that issue is whether there is an essence of a species, be it as a kind or an individual.

I assume in the following that to understand human nature is in the relevant respects independent of solutions to debates about what species, natural kinds, and essences are. In order to stay with the more specific topic of understanding human nature, it is better not to go into the many details of how to spell out the concepts of species (as individual or kind), natural kind, or essence and their relations to each other. I therefore reconstruct the anti-essentialist consensus without a detailed, general account of these categories, even though certain aspects will have to be assumed as being part of the consensus.

The Darwinian Challenge: Two Anti-Essentialist Claims

The Darwinian challenge consists of two anti-essentialist claims. Ever since natural kinds became a topic in twentieth-century philosophy, chemical elements have been regarded as paradigmatic examples of natural kinds—kinds with an essence that is definitional as well as explanatory and, subsequently, descriptive-predictive. The atomic structure of gold, as in the quotation from Ereshefsky above, is a standard example to illustrate such a chemical kind essentialism.

Chemical kinds are, first, spatiotemporally unrestricted classes. Second, there is a (set of) essential characteristic(s) for each kind. In the case of chemical kinds, this is usually taken to be their atomic structure. These essences are intrinsic to the members of the respective kind; they define the kinds (i.e., allow delineation from other kinds and decide kind membership) and explain (at least in part) further properties of the kind members (e.g., the valence and combinatorial relations to other chemical elements). Subsequently, they allow for prediction and description of these further properties. If one knows that a token entity is gold (or knows that it has gold's atomic structure and is therefore gold), then one can not only explain surface properties of the token entity at hand but also predict and therefore describe the properties usually resulting from that structure. Knowing the essence is all one needs to decide kind membership, and knowing a particular entity's kind membership is all one needs for explanation, description, and prediction. This is the epistemic power of essences.

Biological species aren't natural kinds in that sense, claimed biological anti-essentialists like Sober, Hull, and others. First, biological species in the evolutionary sense are spatiotemporally restricted by historical relationships that stem from descent, that is, genealogical relationships. This first anti-essentialist claim has two levels: the level of individual members of a species and the level of species themselves. Species concepts that treat species as evolutionary groups hold that the members of a species need to be in a genealogical relationship with each other in order to belong to the same species. They also hold that species themselves are part of a genealogical branching structure, often depicted as a tree, even though it is probably more like a bush or a network. By contrast, members of chemical kinds need not be genealogically related to be members of the same kind, and the chemical kinds themselves form a periodic table rather than a genealogical system. Take gold again as an example: an instance of gold need not be genealogically related to another instance of gold to be an instance of gold and gold as a kind is not where it is in the periodic table because it descends from its neighboring kinds.

This holds despite the fact that there is cosmic evolution. All chemical elements—except the very first, the lightest element, hydrogen—had to evolve from other chemical elements. Gold, for instance, is a true latecomer whose whereabouts have recently been simulated (Goriely, Bauswein, and Janka 2011). In principle, such an evolutionary story of chemical elements might even contain that gold has evolved (and still evolves) out of platinum, its direct neighbor in the periodic system, or evolved into lead, its other neighbor. But a heavier element does not necessarily evolve later than a lighter one. In addition, there are fusion, decay, and other complex processes. Sulfur, for instance, is not descended from phosphorus (its neighbor) but results, according to contemporary theory, from the fusion of two oxygen nuclei (Gebhardt 2008). As long as gold or sulfur (or whatever one has) would still be classified as they are, according to their atomic structure and independent of their evolutionary whereabouts, chemical kinds are not genealogical kinds. The trajectory of evolution (i.e., the kind's history) is, given current praxis in chemistry, not relevant for where in the periodic table a kind is placed. The current system of ordering chemical elements is logically, historically, and practically independent of the historicity of its elements. Only if the cosmic evolutionary trajectory were to decide which elements are put next to each other in the ordering of chemical kinds (the periodic table), could the identity of the kinds be spatiotemporally restricted by historicity.

Whether one takes historicity as important for classification is a matter that depends on many factors, scientific usefulness being one. Biological species might well one day stop being regarded as historical kinds, and chemical kinds might in the future be regarded as such historical kinds (e.g., if science develops so that it becomes more useful to do so). But currently, the scientific ontology regards biological species as spatiotemporally restricted (i.e., genealogical kinds with historicity) and chemical kinds as spatiotemporally unrestricted kinds.

This also means, as Sober (1980) already stressed, that it is not the fact of evolution as such (and the corresponding vagueness problems at the boundaries of species) that marks biological species off from paradigmatic natural kinds with their essences: "The fact that nitrogen can be changed into oxygen does not in any way show that nitrogen and oxygen lack essences" (356). The same holds for biological species. Hull (1973, 73) similarly noted that the transmutation of chemical kinds and the evolution of species

are very different. Physicists change a small quantity of one element into a like quantity of another element. Lead and gold, defined in terms of their atomic number still remain lead and gold. Even if all the lead in the universe were transmuted into other elements, a spot would still be reserved for lead in the classification of elements. If eventually lead were to form once more, it would be lead, pure and simple. The *process* by which a physical element generated and its past *history* are *irrelevant* to its identity. (emphasis added)

The evolution of species as such (i.e., that they change) is thus not the issue. Rather, the issue is historicity—the dependence of a kind's identity on historical relationships, which stem in the case of biological kinds from the fact of descent—genealogy, in other words. It is the importance of historicity for the identity of the kind that marks biological kinds off from paradigmatic natural kinds with their intrinsic definitional and simultaneously explanatory and predictive essence. The identity of chemical kinds, as understood today, does not depend on history.

But this is just one side of the challenge that population thinking provides. Members of chemical kinds are usually taken to be identical in essential characteristics (e.g., all instances of gold share the exact same microstructure), whereas there is nothing like that for biological species. Variability is everywhere and no essence in sight behind the veil of perceived variation. Life is too messy to follow strict identity. Similarity at the genetic level, the cellular level, the organ level, or the organismic level is all one can use to group individual organisms into kinds. There is then—because of variability—also no cluster of typical and species-specific features of being human that amount to a human nature in the sense of a traditional essence that all and only humans have.

Granted, there is variability in chemistry too. Chemical isotopes are variants of an element that have the same essence (hence the same atomic number) but different atomic weight. Yet as long as the variation is of that shallow kind only, isotopes do not provide an analogue to biological variation. Only if chemical kinds showed similar deep variation as biological kinds would chemical kinds also fail to be natural kinds in the strict essentialist sense. This would even strengthen the claim made here: that variability creates a problem for essentialism.[6]

Variability is thus the core of a second anti-essentialist claim. Variability is the aspect of the Darwinian challenge that gave population thinking its name, when Mayr (1959) introduced the term and contrasted it with typological thinking. The latter regards all individuals of a species as identical. According to Mayr (1976), Jean-Baptiste de Lamarck was such a typologist and believed that although individuals differ in many respects, they do not differ in what counts evolutionarily: they were believed to react in the same way to a particular environment. Imagine the following possible world: there is a population of entities, adapted to an environment that strictly stays the same over time. Each individual, at any time, is strictly the same with respect to all evolutionarily relevant properties. There are perfect copying mechanisms in place (no mutation) for the one species existing (created out of nothing) and there are life and death and therefore biological reproduction. In such a possible world, one would have life (because there is death) rather than mere persistence of entities, but one would not have evolution.[7] If we added to this possible world, that there is change in the environment from within or outside, then the organisms will have to adapt to these changes and hence evolve. But since all individuals are basically the same, the evolution would be transformational. Each individual would adapt similarly and, via inheritance, a transformation of the whole species (rather than just of a few individuals) in the direction of adaptation would result. In such a transformational evolution, as Lewontin (1983) called it, the species changes since each member of the species changes (earlier or later) in the direction of evolution. The crucial point is that although the change happens at the level of individuals, individuals' individuality is irrelevant for the change. Furthermore, in such a transformational pattern, neither variability nor a sorting process is necessary for evolution to take place. Variation and selection are unimportant in such an evolutionary pattern. A comparison to chemical kinds once again can help in clarifying the point: even if chemical kinds showed some variation (e.g., have isotopes) and evolution (as illustrated above), it might still be the case that their evolution happens in a transformational manner. Whether that is the case for chemical kinds is an empirical question that cannot be answered here. The important issue here is that a Darwinian variational evolution, as Lewontin (1983) called it, is decisively different from a transformational evolution.

Variational evolution, the type of evolution that Darwin made prominent, locates evolutionary change at the level of populations and makes variation and a consequent selection process necessary for evolution to occur: individuals are not guaranteed to react adequately and are thus also not guaranteed to react in similar ways to the challenges of the respective environment. The differences between them are not only real but evolutionarily relevant. In such a system, individuals' individuality is not just relevant but absolutely crucial for evolution. Evolutionary change happens by a change in the statistical proportion of individuals, that is, by a change at the level of the population, not by a change in every individual. Consequently, variation and selection are part of all contemporary abstract recipes of Darwinian evolution (from Lewontin 1970 to Godfrey-Smith 2009).

Furthermore, a transformational evolution is necessarily developmental: without development, there would be no evolution in such a picture since development, in an abstract sense, simply is change in individuals. A variational evolution, however, would be possible for entities that do not develop, since variational evolution only needs recurrent variation (e.g., by mutation) and a sorting process.

Although a variational evolution is not dependent on the development of individuals (i.e., change in individuals), such a populational picture is deeply individualistic, but in a different sense from a transformational, typological picture. In a populational picture, the source of change is individuals' individuality. For Lamarck as well as for Darwin, individuals are reformers, but for Lamarck, they are so because of their lack in individuality; for Darwin, they are so precisely because of their individuality.[8]

An essentialist can certainly take variation between individuals to be real, but she would argue that even if members of a natural kind might differ in many respects, they do not differ in their essential properties. That means that essentialists use ways to discount observed variation. Population thinking can then be taken to object to any essentialist variation-erasing procedures: because of the quintessential variability of all life, any variation erasing distorts what is going on. Recourse to hidden capacities in order to arrive at something shared, despite the apparent variation among individuals, is thus in tension with Darwinian population thinking. (Such recourse to capacities will concern us in more detail in section 3.3.)

Finally, to be exactly like paradigmatic natural kinds, the definitional essence (the necessary and sufficient conditions for species membership) would have to be one and the same thing as the explanatory essence—the properties that allow the discounting of variation because they explain why things are similar. For instance, our faculty of reason (or rationality) has often been picked out as essential for the kind of animal we are: it defines what it means to be a human animal, decides about which entity belongs to that biological kind, and allows explaining, predicting, and describing a lot of other traits that are specific and typical in humans, among them the social interactions among humans and the inventions and creations of humans, from architecture to art, and so on. But as I argue in the next sections, neither rationality nor any other potential qualitative trait or combination of traits of the human life form are, first, individually necessary and jointly sufficient conditions for being a member of the species *H. sapiens* and are, second, at the same time explanatory for the human life form.[9]

In the following sections, I discuss the two anti-essentialist claims, anchored in genealogy and variability, by focusing on two of the epistemic roles separately: the classificatory and the explanatory. The third epistemic role, the descriptive-predictive one, is less central for the discussion of whether biological species have essences; it will, however, be of concern in part II.

3.2 Challenging the Classificatory Role of Essences

Imagine a list of species-typical and species-specific traits: how humans move (bipedalism), how humans communicate (human language), how humans think (rationality, morality), and so on. These are the behaviors or properties that a comprehensive description of how humans are like might note. It lists traits of humanness—the life form of the human species.[10]

To fulfill the classificatory role, that is, to successfully include all humans (past and recent) and exclude all (past and recent) nonhumans, the essentialist has to refer to at least one such trait, more likely a list of such traits, as necessary and (collectively) sufficient conditions. And here is the rub: there are no necessary and collectively sufficient conditions for membership in *H. sapiens*. First, given ethical discussions about disabled humans, it seems to be sufficient to be genealogically related to *H. sapiens* to count as human. There can then always be a *H. sapiens* (i.e., an entity conceived by *H. sapiens*) who does not fulfill one, some, or any of the conditions on the list of traits belonging to humanness. That means no intrinsic, qualitative property (like rationality or the like) is necessary to be a member of *H. sapiens*. Second, there can always be a humanoid, an entity that fulfills all the humanness conditions (i.e., exhibits all the traits of the human life form) who is not a human in the biological sense. This means that humanness is not sufficient to be a human in the biological sense. Let me add some details to these two points about necessity and sufficiency of qualitative traits.

Qualitative Traits Are Not Necessary

As said, there can always be a *H. sapiens* (i.e., an entity conceived by *H. sapiens*) who does not fulfill one or more of the qualitative conditions on the list for humanness and would nonetheless be regarded as a human. That means that the respective qualitative traits that this individual might miss are not necessary to be a human. Consider a disabled person: a person of severely restricted growth, a person who does not walk like most other humans but moves in a wheelchair, a deaf person who cannot communicate like most other humans.[11] There is no doubt that these people are human beings. They are as much members of *H. sapiens* as any nondisabled people.

The status of extreme cases of nonviable fetuses, by contrast, is contested in many senses, but there are at least two classificatory contexts in which they are clearly counted as members of the human species. Take, for example, the breathtaking fetuses in the glass cases of museums of history of medicine. These extreme cases are considered human beings, despite the fact that they (if alive) would have missed literally all of the typical and specifically human characteristics. Interestingly, in the Berlin Museum of Medical History at the Charité, they are in many cases not labeled as human but simply as fetus. It seems that their status as humans is not challenged and needs no extra mentioning. For the completely amorphous fetus, depicted in figure 3.1, the label contains explicit reference to the human species and says *monstrum humanum amorphi*.

In terms of similarity (i.e., the sharing of qualitative traits), this creature is as far from any living human being (or the respective normal human fetus) as anything can get, but it is considered and labeled as human nonetheless. And, I reckon, in that case, the label was necessary since otherwise the spectator would have had literally no other indicator of the human nature of that creature.

To regard nonviable results of human reproduction as members of the species *H. sapiens* is also in accord with the head-counting practice in evolutionary biology, as Sober (1984, 339) stressed:

> In the theory of natural selection, one considers the life-history of a population as extending from the egg stage to the adult stage, with selection potentially acting anywhere along the way. Zygote inviability is a kind of mortality selection; it just happens early. Zygotes are in the population. And if they're in the population, they're in the species. ... Whatever may be said against counting fetuses as *persons*, I

The Darwinian Challenge

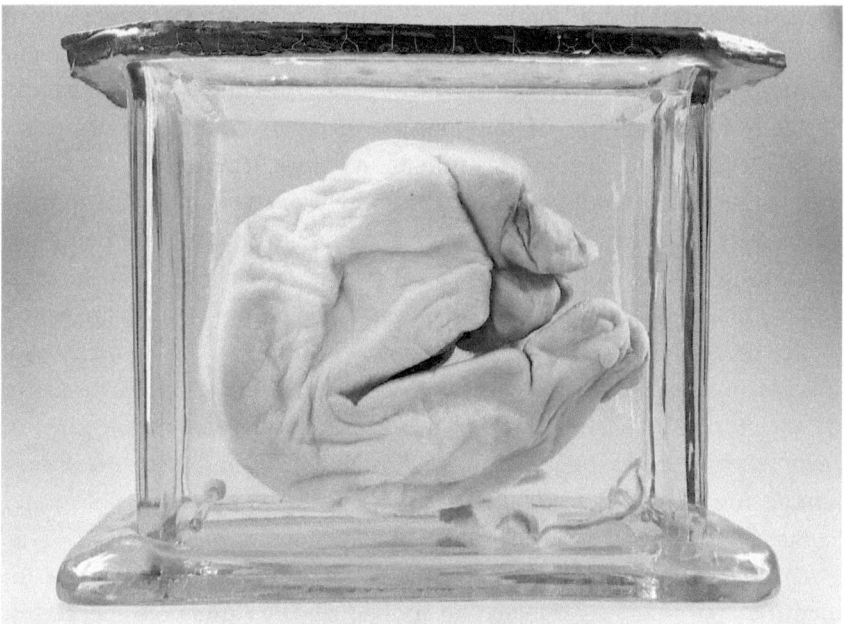

Figure 3.1
Nonviable human fetus. This fetus has the label *Monstrum humanum amorphum*. Without the label, it would be hard to recognize its species membership because of its amorphous development. (Collection: Berliner Medizinhistorisches Museum der Charité, end of nineteenth century. Photograph: Navena Widulin. Printed with permission by the museum.)

see no biological motivation for denying that they are *organisms* in the same species as their parents. They are members of *Homo sapiens*, and are human beings in that sense. (emphasis in the original)

Thus, also given the classificatory practices mentioned, it is sufficient to be genealogically related to other *H. sapiens* to count as a human in that sense.

Qualitative Traits Are Not Sufficient

The set of qualitative traits typical (or specific) for the species not only fails to be necessary to be a *H. sapiens*; it also fails to be sufficient. After all, there can always be a humanoid that is not a human in the biological sense. As long as one agrees that there are reasons why one calls that entity a humanoid rather than a human or *H. sapiens*, it holds that intrinsic, qualitative traits such as rationality and language are not sufficient to be a member of the species *H. sapiens*. The humanoid can certainly still be regarded as a full person but is not therefore a member of *H. sapiens*. Thus, one can certainly decide (via stipulation) to equate human with person, but I do not think this matches the vernacular use of the term *human* or *person*.

Imagine what philosophers call (since Davidson) a swamp man: an individual that is an exact replica of a specific human, coming into being out of nothing, that is, out of a miraculous swamp, without any genealogical or historical relationship to other humans, including the human whose replica the individual is. It is a perfect out-of-nothing replica. To make it so, it is best to synthesize the swamp at and with material from a twin earth, the exact hypothetical replica of our beloved Earth. There are a couple of these swamp men in the literature on human nature, usually copies of (in-)famous men. Hull (1978) thus had a twin Hitler to make the point, and Samuels (2012) had a twin Obama.

Given a concept of species that takes genealogy as indispensable for identity, one might regard such an individual as a humanoid (because it moves like humans, talks like humans, thinks like humans), but one will not regard it as a human in the biological sense of being a member of the species *H. sapiens*. One would, however, as the standard assumption among philosophers goes, count an exact twin Earth replica of a sample of the natural kind gold as gold, even if it had no genealogical relationship to any instance of gold on earth. This thought example is meant as evidence that we, if pressed, indeed distinguish between human and person and that we do so by connecting the first to genealogical historicity. Genealogy is important in our self-understanding as humans.

At the level of species (rather than the level of individuals partaking in a species) a similar case can be made, as Okasha (2002, 196) stresses: "Sibling species are morphologically indistinguishable (or very nearly so) but treated as distinct because they form separate reproductive communities—they engage in little or no interbreeding." And as Rieppel (2013, 165) states,

If a species is lost through extinction, that extinction is terminal: were an "identical" (not numerically identical, but phenotypically and/or genotypically identical) species to re-evolve, it would still not be the same species as it would have a different evolutionary origin ... or history.[12]

Whether it is the species level or the level of individuals partaking in a species, genealogy trumps similarity. That genealogy trumps similarity for species delineation as well as membership is suggested and justified by reference to current phylogenetic praxis (I further justify the importance of genealogy in chapter 5). In addition, it also has a foundation in another (and maybe even independent) source: our prejudicial attitude toward "our stock." Thus, even if taxonomic praxis in evolutionary biology were to change and started giving priority to similarity over history, the argument that genealogy trumps similarity in our way of understanding being human might still have a social foundation, a point I come back to in chapter 5.

Any full argument with respect to these two points about qualitative traits of the human life form being neither necessary nor sufficient for membership in the kind certainly would have to involve a proper empirical study on use of the terms *human* and *person* in different social contexts, as well as more detailed case studies on a broader diversity of medical and scientific contexts. It is expectable that there is variation in usage. Yet I do not know of any such comprehensive studies and can thus proceed only on the assumption that the intuitions about disabled people and humanoids used here for contemporary society and science are correct, at least in the context of what is still often called the West.

The Problem of Squaring the Circles

Graphically depicted, variability leads to an impossibility of squaring the circles, as depicted in figure 3.2.

Recall figure 1.1, where the square (H) represented the members of the group *H. sapiens* genealogically defined. The boundaries are allowed to be fuzzy since at the beginning and end of the species, there will be borderline cases of individuals for whom it is unclear whether they (already or still) are *H. sapiens*. This is a simple consequence of the fact of evolution. But that is not the important issue. Important is that none of the qualitative traits (e.g., reason, language, shared intentionality, tool use, or the opposable thumb) that have been discussed over the centuries for being the essence of our species includes all possible humans and only humans. If one picks one species-specific and species-typical trait, such as language in a narrow sense (represented by the circle L), there can always be some individuals who are clearly humans in the biological sense who nonetheless do not have that trait, for example, if they have an articulation difficulty involving grammatical impairment, as reported in Enard et al. (2002); thus, the

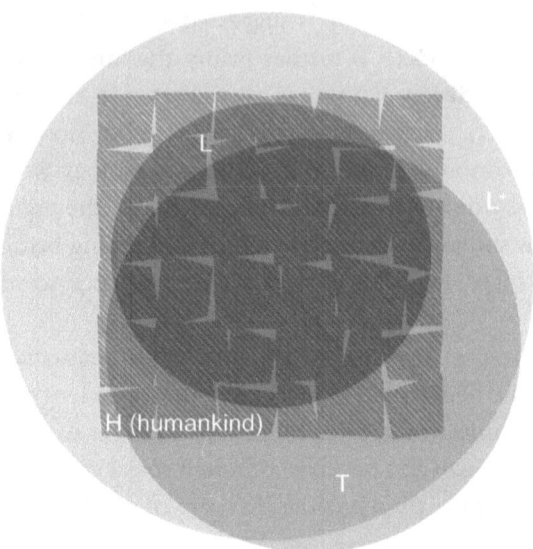

Figure 3.2
Squaring the circles. The square represents the set of all *H. sapiens* (H). The circles represent qualitative features of the human life form that have been considered as deciding who counts: L (for language narrowly defined), L+ for language broadly defined, and T (for tool use). No single feature and no combination of features will ever match exactly the square. (Title of the figure inspired by Griffiths 1999. Illustration: Vera Brüggemann, Bielefeld.)

trait cannot be necessary for species membership. If one therefore tries to be more inclusive and goes for a broader definition of language (L+) so that these humans, who nonetheless have some language ability, are exhibiting the trait "language," then one gets a different problem: one will include entities that are not members of *H*, such as a humanoid or an animal that communicates not exactly like a human but still similarly enough to say these animals communicate using a language. Consider the hypothetical ape that La Mettrie in *L'homme machine* ([1748] 1912, 31, 103) used as his thought experiment, who becomes "un homme parfait, un petit homme de ville" (a perfect man, a little gentleman) by learning to communicate like humans since it got taught sign language. In current philosophical literature, that "homme parfait" has been revamped as the "superchimp" (McMahan 2002).[13] Or consider any of the signing apes that contemporary primatology refers to. Exhibiting L+ (that broadly defined trait) can then not be sufficient to be a human since these animals exhibit that trait but are not therefore considered humans.

From this, the first important point derives: for any given qualitative trait of the human life form, the trait will fail to be at the same time necessary and sufficient for membership since there is a dilemma: the selected property is either too broad (failure in sufficiency) or too narrow (failure in necessity). When we defined language broadly, animals of all kinds exhibit it, and the trait fails in sufficiency. When we defined language more narrowly, to prevent the inclusion of speaking nonhuman animals, then the properties' intension changes and the extension gets narrower (i.e., the circle gets smaller), but then, in treating that trait as necessary, one also excludes more humans. Failure in necessity results. More technically, if one fixes the problem at the side of sufficiency (moving to a property that is not too broad so that having the trait is not allowing nonhuman animals in), then one gets a problem at the side of necessity (one excludes many humans too, since they also will not have the trait).

The second problem is that if one continues with that game (imaging different circles like L), one simply gets further circles, with yet another extension, for example, T for tool use. In the end, one ends up with the following: each trait has a different extension than the other traits do and none fits exactly H. In other words, even if we get a cluster of circles, a property cluster, nothing will ever square the circles.

Only the square will match the square, that is, only the relation "to be genealogically related to other humans" will match it. Evidently it matches H only since that is how one has set up H. In the language of necessary and sufficient conditions, this genealogical nexus is the only property (if we count relations as a property, for the sake of the argument) that is necessary and sufficient for species membership (except at the vague beginnings and endings of the species). For Hull, this was it. Human nature, if taken to refer to those conditions that "make us human" in the classificatory sense (i.e., conditions that decide species membership), then human nature is nothing but a genealogical nexus.

To do so, however, seems to be trivial since it does not (at least not at first glance) do any explanatory work, as Okasha (2002, 203) stressed. Recall that traditionally, essences were conceived as definitional as well as explanatory. Pointing to the genealogical nexus to answer what it means to be human is thus at first glance close to a tautology. Who is human? Those born to (or at least conceived by) humans. Despite the triviality involved, the property does provide what Hull was concerned with: it is useful for one specific epistemic role of the concept of a species' nature, namely, the classificatory role at the level of individuals who are grouped into a species. He argued that, in all practically relevant situations, the genealogical nexus successfully helps to decide whether an individual belongs to the human species. He stated that this claim

> is not negated by the fact that species periodically split or bud off additional species. To the extent that speciation is "punctuational," such periods will be short and involve only a relatively few organisms, but inherent in species as genealogical entities is the existence of periods during which particular organisms do not belong unequivocally to one species or another. *Homo sapiens* currently is not undergoing one of these periods. The genealogical boundaries of our species are extremely sharp. The comparable boundaries in character space are a good deal fuzzier. (Hull 1986, 4)

What Hull called character space, I call humanness or the human life form. It follows, if all of the above is correct, that the species that is usually meant when people talk about human nature is genealogically defined: it is necessary and sufficient to be conceived by a human to be a human.

If that were all the concept of human nature was for (to exclude entities from the in-group of humanity and to include all kindred ones), then such a minimal classificatory genealogical concept would be all one needs. Given that parsimony is of value for epistemic endeavors, one should not postulate more than what one needs. Consequently, one should stop the reflection about human nature right here. Yet deciding species membership was not the only thing for which the concept of a human nature was used, as Roughley (2011) already objected to Hull. But before I analyze the other epistemic roles of the concept in sciences, let me take stock.

The swamp man lacks the one necessary and solely sufficient property that all and only humans must have for being *H. sapiens*: the genealogical property of being genealogically related to other human beings. How the genealogical nexus is related to the human life form (i.e., descriptive nature, humanness, life form, character covariation, as Hull called it— choose your preferred term), and to an alleged explanatory nature, will be spelled out in part II. Most important in this chapter are the following points. First, the genealogical nexus is not an intrinsic, qualitative property; it is not a trait of the human life form. Second, it can be used only

The Darwinian Challenge 49

to identify members of a species in situations in which there is already a human species. Without there being a human species, the classificatory question (whether this or that entity is a human or not) would simply not make sense. Third, although genealogy is often unobservable, it can still function as the necessary and sufficient logical criterion for species membership, for example, by having observable surface properties as epistemic indicators for it. Fourth, we (those having a discourse about humans and human nature) might nonetheless accept some entities that are humanoid to be humans in a different, namely, a social (i.e., nonbiological) sense, as Hull (1978) also admitted. If an entity is humanoid and interacts with *H. sapiens* socially (rather than biologically), that is, acts as a member of a social or ecological group, that individual might well be regarded as a human in that social sense, albeit without becoming thereby a member of the species *H. sapiens*. From the dehumanization challenge it is clear that this social concept of being human is (as its biological counterpart) first and foremost classificatory, defining the boundaries of a group, that is, who is in and who is out.

3.3 Challenging the Explanatory Role of Essences

Around the nineteenth century, not only an essentialist way of classifying people got lost, but also an essentialist way of explaining what makes us human. Given that biological variability is simply an observational fact, the shift from an essentialist to a populationist picture is in how observed variation is explained.

There are different names for what an essentialist way of accounting for biological variation amounts to. For Mayr (1959), it is part of typological thinking; Sober (1980) called it "natural state model" and Matthews (1986) "norm-defect theory." Lennox (1987) talks about a "teleological essentialism" (340) based on "teleological explanation" (357) of differences between and within species; Walsh (2006) also talks about a "teleological essentialism." Aristotle is the alleged historical figure cited as having defended such an explanatory essentialism.[14] I briefly described it with respect to the dehumanization of women in chapter 2. In this section, I characterize the alleged contrast to population thinking without getting drowned in the pitfalls of interpreting Aristotle's philosophy in a historically adequate manner for which I lack space as well as expertise.

Independent of historical issues, it is clear that for Aristotle, women and slaves were members of the same kind as Greek men, but they were nonetheless deviations from human nature. This involved reference to genealogy and to a teleological essence. I will say a few more words on genealogy and then focus on essence, which is the important aspect for understanding the differences in how an essentialist and a populationist explain variation.

There are two ways to interpret why Aristotle counted women and slaves as members of the same species as Greek men and to accommodate this with his explanatory essentialism. First, genealogy might have already been important in Aristotle for identifying species boundaries and membership. We would then have to interpret Aristotle as using the concept of form as essence (formal nature), not for classificatory purposes, so that from the classificatory point of view, an individual is a human because of its genealogical relationship to other humans, while from the point of view of the teleological essence, that very human being is not fully partaking in human nature.[15] Alternatively, one could read Aristotle as having children and deviant people as belonging to the same species as men because they at least have the capacity (i.e., the potential to, or the disposition for—choose your preferred term) to develop the natural state.[16] An individual who turned out female or slave just had bad luck; that individual, however, can still take comfort from being attributed at least a potential, a capacity for the full human form, and therefore membership in the kind. Only in this interpretation would the classificatory and the explanatory role be fulfilled by the same thing, the form of the human species. In the first interpretation, only the explanatory role is fulfilled by that essence.

I take it to be a matter of historical intricacy how Aristotle is best interpreted. I will not take a stance since the difference between the two options is not important in this study. What is important with respect to the challenge of the explanatory role is that the teleology involved in Aristotle's thinking was essentialist. First, he picked out some qualitative traits (such as rationality) as essential and variation discounting, contrasted with others deemed to be negligible for the essence of being human. Second, this involved attributing more or less of the explanatory essence to particular groups of individual humans, leading to dehumanization. I analyzed that connection between understanding human nature and dehumanization in chapter 2. Here I continue with illustrating how such an essentialism connects to the explanation of variation and how population thinking challenges the postulation of any normalizing capacities, taken as teleological essence of being human. The exploration will be continued in chapter 8 with a discussion of two proposals to save something like an intrinsic explanatory essence, despite the Darwinian challenge.

Here is a simple example. Figure 3.3 has a hypothetical bell-curve-shaped so-called normal distribution for human body height. It depicts a hypothesized variation of human body height. Since all one can literally see is the pattern of variation, natures in the sense of explanatory essences must be hidden (if they exist). They are theoretical, that is, unobservable entities. In that sense, an essentialist is not denying that variation exists. As long as there is an essence behind it, variation on the surface is irrelevant for an essentialist. For the example of body height, the hidden explanatory essence would be a capacity for a specific range of human body height that explains body height variation, for example, the capacity of humans to reach a body height of roughly between 150 and 190 cm. Another example would be the

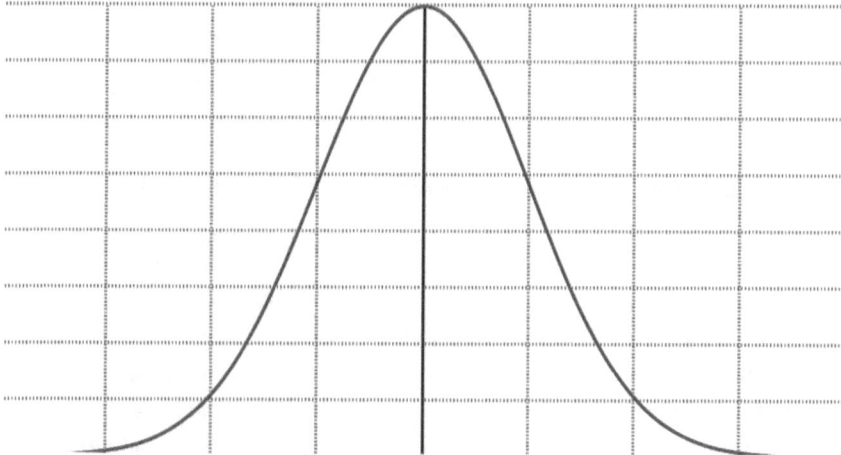

Figure 3.3
What variation looks like. A hypothetical normal distribution of a property, depicting how many individuals (vertical axis) have a certain property (horizontal axis). Take body height as an example and imagine the mean (the vertical line in the middle) to be 170 cm body height. Most humans will then be between 150 and 190 cm tall if the property has a normal distribution. (Illustration: Stefan Pohl; adapted and used, according to Creative Commons CC0 1.0 Universal Public Domain Dedication.)

capacity for human spoken language: claiming that it exists makes everybody equal in terms of that fundamental trait of being human, which is often postulated as existing in every human (just by being a human), even though not all and not only humans express that trait.

Such traditional hidden explanatory essences are meant to neutralize the variation: if there were such capacities that are truly shared by all members of the species *H. sapiens*, then one would not have to worry about the variation in the respective trait. Variation would play no epistemically important role. It would be irrelevant since it can be discounted because of the existence of a hidden explanatory essence, an essence that proverbially lets you have your cake and eat it too: the explanatory essence explains normal outcomes (if proper triggering causes dominates) as well as abnormal ones (if disturbing triggering causes are present).

The problem with such an explanatory essentialist schema is that current evolutionary biology has no need for it, and it might even be incompatible with Darwinian ontology. In the following, I defend the dispensability argument that essences are dispensable for explanatory purposes within a Darwinian frame, then I discuss why they can even be taken to be incompatible with a Darwinian ontology. I finally claim that all that is left is propensities of kinds, which I take to be distinct from capacities of individuals.

The Dispensability Argument: Darwinian Theory Has No Need for an Explanatory Essentialist Schema

It was, according to Sober (1980), Darwin's cousin Galton who introduced a truly nonessentialist, populational way of accounting for variation: an approach that accounted for variation via heredity rather than normalizing explanatory essences. Variation in a current generation is explained by the combination of the following: variation in a previous generation, newly added variation from mutation or developmental plasticity, and the laws of heredity. Period.

There is no need to account for the range of variation, the statistical pattern observed, by reference to a hidden normalizing capacity (or any other kind of explanatory essence), since the theory of heredity together with development provides an alternative sufficient explanation that is in no sense inferior to an essentialist explanation. It is, rather, the other way around: the essentialist explanation would be a superficial addition.

The Darwinian Challenge 53

Consequently, there is also no need to distinguish between normal "proper" developmental resources and disturbing "improper" ones. Variation is all there is. Thus, the new populational way of accounting for variation establishes not only parity between individuals (if everybody is different, there is no "less human" anymore), but also parity between causes: there is no epistemic use for a distinction between normal and disturbing causal factors that explains why there is variation. This argument, that Darwinian theory has no need for an explanatory essentialist schema, is clearly a dispensability argument.

The Incompatibility Argument: The Darwinian Theory Has No Room for Explanatory Essences

An essentialist might reply that despite dispensability, capacities can still be compatible with Darwinian theory, since dispensability does not imply incompatibility.[17] Correct. Yet compatibility of explanatory essentialism with population thinking can also be challenged. Hull (1986) and Sober (1980) already suggested that because variation is essential to biological species, there is not only no need, but also no room for an explanatory essentialist schema. Since they never spelled out in detail in which sense there is an incompatibility, I will do so here.

First, since it is necessary within Darwinian theory that there is variation (so that selection has something to act on), a unit of selection needs to vary to be a unit of selection, but essences do not vary. Thus, since essences cannot be units of selection and be explanatorily important in evolution, they are incompatible with evolutionary thinking. This can be described as a dilemma: either explanatory essences vary too (so that behind the overt variation, there are also individual differences with respect to essential capacities), so that they can be units of selection, but then they do not help to get rid of the variation between individuals; or explanatory essences are pure epiphenomena (i.e., no units of selection) and thus do not need to vary since they are—as epiphenomena—not themselves causally involved in the evolutionary business. The dilemma is this: essences cannot be explanatory and variation reducing at the same time.

Second, the essentialist lacks an explanation for a certain kind of parity of causes that is part and parcel of evolutionary thinking: an explanatory essentialist would have to show why what she calls disturbing factors occurs (and has to occur) regularly for evolution to happen, because otherwise there would be no variation. This regularity of disturbing factors was already a problem for Aristotle, who, on the one hand, believed that women have to occur regularly (for there to be reproduction and any further species members), but, on the other hand, claimed that they are nonetheless not on a par with men, who are the "normal" (i.e., ideal) realization of the essential form of any sexually reproducing biological species. I do not see how the essentialist has the resources to answer this problem, whereas the population thinker does not need to answer it since it has the parity established from the start by not distinguishing between disturbing and nondisturbing factors.

There is no space here to develop these two points further, but they should be enough to cast doubt on the compatibility of explanatory essences with Darwinian ontology. For the purpose of this study, the dispensability would already have been sufficient to move on to what is left if essences are not part of the scientific picture of human nature anymore. Yet before moving on, I point to a third problem that finally allows reconstructing the alleged variation-discounting capacities of the essentialist as what I would like to call *population-level kind propensities*.

Why Tigers Have a Population-Level Kind Propensity to Develop Stripes

If variation-negating explanatory essences existed, they would also be too unlimited, in the sense that their application could not be constrained to one species. This is how Hull (1986) spelled out his incompatibility claim against explanatory essentialism. His point was that if (to save the essentialist picture) a human is said to nonetheless have a capacity for trait F (qua being human and in contrast to a nonhuman animal), despite evidently not exhibiting F, then other creatures that are not members of the species at issue also can be said—and on the same metaphysical basis—to have the respective capacity. Essences would then fail to be specific for a species.

I develop his argument by using the kind of typical toy thought experiment that essentialists themselves like to use. I switch from humans to tigers to prevent interference of the normative issues related to being human. In addition, tigers with their stripes are a stock example of the essentialist discourse. Imagine a mutation that is rare for tigers and happened to unfortunate little Tiggy, a little baby tiger. The mutation is causally important for explaining why Tiggy will not develop stripes, unlike most other tigers. I'll call the allele for stripes STRIPY and the mutation Tiggy has ASTRIPY.

Because of ASTRIPY, Tiggy will be stripeless, and because of that, she will not be able to participate in the tiger-specific life form. Imagine that having stripes is really important for being a tiger, for example, for getting tiger friends. Nonetheless, Tiggy is a member of tigerhood since—the essentialist might answer—she has at least the capacity to develop stripes—simply by being a tiger.

Now imagine a baby lion, Leo. Leo does not have stripes either. Can one say that Leo has the capacity to develop stripes? Sure, if Leo had STRIPY (and, if necessary, further respective mutations, irrespective of how many), he would develop stripes. We only need to assume that the only thing that is making the difference for having stripes or not is having STRIPY or ASTRIPY; all other genes that are relevant for the production of stripes happen to be the same in Tiggy and Leo, which is not inconceivable given that lions and tigers are genealogically quite close.

What prevents one from saying in such a possible world, "If there were not this 'interfering cause' [the allele ASTRIPY] with Leo's capacity for stripes, Leo would have had stripes," as Tiggy would have had stripes if she had STRIPY. In other words, why would one still not talk about "interfering causes" in the case of Leo, whereas one would talk that way in the case of Tiggy? Why is Tiggy, the baby tiger, treated differently from Leo, the baby lion, even though intrinsically they are the same? STRIPY is missing in both, in Tiggy as well as in Leo. Why are the same missing causes in the one case interfering causes and in the other case not? Because whether something is an interfering cause or not depends on the reference class; that is, it depends on what is normal, statistically speaking. And what is normal depends on whether one puts Tiggy or Leo into the group of tigers or lions.

Putting an individual into a kind is thus carrying information about what to expect from such an individual in addition to being a member of the kind. This is how the descriptive-predictive role of essences (introduced in section 3.1) connects with the classificatory role. Once an individual is put under tigerkind, stripes are predicted (i.e., expected). Once an individual is put under lionkind, no stripes are expected. This is the reason one treats Tiggy differently from Leo. As individuals, Tiggy and Leo are the same with respect to their capacity of having stripes.

It follows that if the actual developmental resource (ASTRIPY) is atypical of a particular sort of being (e.g., a being descendant from tigers such as Tiggy), then the allele for ASTRIPY is an interfering cause. It gets foregrounded since it is what is abnormal. If, by contrast, the actual developmental resource (ASTRIPY) is typical of a particular sort of being (e.g., a being descendant from lions, such as Leo), then the allele for ASTRIPY is not treated as an interfering cause since it is what is normal for that reference class.

Thus, another dilemma for the essentialists results: either the capacity to develop stripes is also present in species where the stripes are rare (or even not existing at all), but then capacities are too frivolous; or the alleged capacity to develop stripes is not an intrinsic property but depends on whether the entity counts as tiger or lion. If the second horn of the dilemma is chosen, the capacity fails to be an intrinsic property.

Yet there certainly are population-level kind propensities, for example, the propensity of tigerhood (the species as a whole) to produce individuals with stripes frequently. But this is a propensity of the respective populations, not a capacity in individuals making up the population.

A final point is that genetic factors, best candidates for causally essential properties for those believing in them as master molecules (e.g., as Antony 1998 still does), will not come to the rescue. They not only turned out to be quite dependent on other developmental resources (as I illustrate in chapter 4), they also turned out to be so diversely distributed that they hardly provide comfort for explanatory essentialists. Of the 1 percent of genetic material in humans that is specific for humans, 0 percent is shared by all humans.[18]

In sum, the contemporary biological model to account for variation is statistical through and through, down to the genetic level; explanation is consequently populational (oriented toward statistical traits of a population) rather than typological (oriented toward types, that is, ideal, normal, representative individuals instantiating the essence).

This has five tenets: first, variation is no longer "something to be explained or explained away," as Sober (1980, 370) stated, but it is something that is itself explanatory; it explains later variation. Second, there is no need in evolutionary theory to point to essential (i.e., hidden, microstructural, constitutional, internal) properties that explain (away) variation as the result of different kinds of realization or instantiation conditions, disturbing or nondisturbing factors. Third, in contemporary Darwinian ontology, there is not even room for capacities behind the variation since using them leads to the dilemma that they cannot be simultaneously explanatory and variation discounting; fourth, if there were such essential capacities, they would be too frivolous, too widespread, applying to entities across the species boundaries, entities that should not have these capacities, given the setup of the essentialist reasoning. Fifth, if at all, then there are kind propensities—the propensity that certain developmental resources are present for this or that individual of a specific kind. That propensity is, however, a property of the population. The individual has either this or that set of developmental resources.

Thus, Darwinian ontology opposes a typological way of explaining the human or any other life form. Since population thinking constitutes an alternative way of explaining variation, the issue is not only a metaphysical one but an epistemic one: a typological way of explaining variation is not just ignoring variation but epistemologically mistreating variation.[19]

In chapter 8, I discuss some replies to the Darwinian challenge, in particular, the teleological natures that Walsh (2006) suggested and the intrinsic essential properties that Devitt (2008) defended. I will claim that they are not populational enough to meet the Darwinian challenge. Nevertheless, I will argue, there is something that can take over the explanatory role of explanatory essences—something like an explanatory essence, something that helps fulfill the traditional role of explaining (at least in part) humanness. I will show that this nonessentialist explanatory nature explains developmentally why the properties of a species' life form occur in individuals of the species in stable, predictable ways and, evolutionarily, why these properties cluster.

3.4 Situating the Anti-Essentialist Consensus

Essentialism might be a viable option for contemporary metaphysics (as in Ellis 2001 or Bird 2007) for reasons that are independent of the issues discussed in this study. Essentialist thinking might also be widespread in humans interpreting other humans in daily life, as psychologists claim under the heading of folk or psychological essentialism (e.g., Gelman 2003; Linquist et al. 2011). Both kinds of essentialism are of no concern here since metaphysical and psychological essentialism have no bearing on the Darwinian challenge. They do not infringe the consensus in philosophy of science, which is that the anti-essentialists are right: an essentialist concept of a human nature (or any other species' nature) is not needed and not compatible with the contemporary biological knowledge regarding evolution, heredity, and development.[20] The Darwinian challenge exists: it says that human nature in the sense of a classificatory or explanatory essence of *H. sapiens* does not exist. Nature as essence is a mirage of an outdated biological ontology—a way of speaking, at its best, that has no grounding in scientific ontology.

This result matches a tradition of critique of essentialism in other philosophical areas, such as gender or disability studies, where essentialism (via the normalcy considerations entering it) is regarded as a main mechanism for dehumanization. Belief in variation-discounting explanatory essences makes it possible to regard some people as realizing the essence and some as failing to do so. If essentialism cannot be used, dehumanization is likely to weaken since variation is then taken as deep and as explanatory, as not to be discounted. Thus, taken together, the Darwinian challenge and the dehumanization challenge point at elimination of any essentialist version of the concept of human nature. The next chapter introduces a challenge that points in the same direction, although it is only indirectly (if at all) related to the kind of essentialism discussed in this chapter. Rather, it concerns the contrast between nature and nurture, a contrast that is often used in a sense that violates the parity of causal factors mentioned briefly in this chapter.

4 The Developmentalist Challenge

Reference to human nature (in science as well as society) often includes not only that the respective nature is an essence (with its multiple epistemic role) but also that nature is to be distinguished from nurture, which includes culture.[1] Conceptualizing nature as distinct from nurture involves locutions such as "X is due to human nature." It treats human nature as an explanatory category. The developmentalist challenge, the topic of this chapter, is questioning the distinction between nature and nurture. I take the core of the challenge to be that biologically inherited and other causally important developmental resources are too entangled to regard them as separate kinds of causes with apportioned kinds of effects—nature on the one hand, nurture on the other hand. Keller (2010) provides a forceful recent elaboration on that challenge.

The developmentalist challenge is independent of the Darwinian challenge. For Schaffner (1998), from whom I take the term *developmentalist challenge*, it is about the future of ideas about genetic causation and DNA primacy. Genes are often at the center of so-called nature-nurture debates since they have often been identified with the essentialist explanatory nature and contrasted with the rest—all the other causal factors influencing development, summarized under the label "nurture." So-called genetic determinism, genetic reductionism, genetic essentialism, and gene centrism are slightly different ways of claiming an explanatory primacy of genetic factors. They are positions that have been forcefully attacked. Developmental systems theory has been most outspoken about the developmentalist challenge, with Oyama (1985, 2000) as a founding figure and others who have contributed in various forms (e.g., Griffiths and Gray 1994, 2001, 2004).

I analyze in section 4.1 how *nature* and *nurture* became historically conceptualized as separate kinds of causes with apportioned developmental effects. Then I introduce in section 4.2 what is considered problematic about the nature-nurture divide at the developmental, intergenerational (epigenetics), and the evolutionary levels (gene-culture coevolution, niche construction). Section 4.3 analyzes in detail the so-called interactionist consensus, about nature and nurture interacting. On that basis, it is possible to be more precise, in section 4.4, about what the developmentalist challenge actually amounts to. The ultimate aim of discussing the developmentalist challenge is, as with the other two challenges, to clear the ground for an account of human nature that is not only post-essentialist and pluralist but also interactive with respect to the involved nature-nurture divide.

4.1 From Physis versus Nomos to Nature versus Nurture

Physis/Nomos in Greek Antiquity

As the origin of the concept of essences, the origin of the contrast between *physis* (nature) and *nomos* (custom, conventions, law, culture) is taken to be in Greek antiquity. For the sake of the argument, I assume that as part of that contrast, physis represented what is now called *nature*—something physical. Nomos, by contrast, represented those things that are relative to cultures and culturally transmitted. The Hippocratics, for instance in the treatise *On Airs, Waters, and Places*, used the distinction between nature, culture, and environment to explain commonalities and differences between groups of people and their characters. So even at that time, there is a nature-nurture divide: nature and nurture (culture and environment) are distinguished as kinds of causal influences for the development of properties of organisms.

Yet, viewed with hindsight, the contrast between nature and nurture was not yet fully developed in Greek antiquity. The main reason is that the Hippocratic authors (e.g., in *Airs, Waters, and Places*) assumed an inheritance of acquired characteristics, which was—historically unfortunate—later called Lamarckian inheritance. If acquired characters can become heritable, as it was imagined in Greek antiquity, then nomos successively turns into physis. As a consequence, nomos and physis become quite entangled since they change together. According to the historian Heinimann (1945), the whole

The Developmentalist Challenge 61

Hippocratic corpus was full of inheritance of acquired characteristics, and it was common sense and stayed so well into the nineteenth century.[2] It was challenged only toward the end of that century. As a consequence, the divide between physis and nomos hardened only then and was generalized as nature versus nurture.

The Advent of Heredity and the Rally against Lamarckian Inheritance

According to the received historical view, Charles Darwin's cousin Francis Galton introduced the modern nature-nurture divide. In 1874, in his *English Men of Science*, he famously introduced "the phrase 'nature and nurture'" as

> a convenient jingle of words, for it separates under two distinct heads the innumerable elements of which personality is composed. Nature is all that a man brings with himself into the world; nurture is every influence from without that affects him after his birth. (Galton 1874, 12)

Galton's divide was one between *nature* as referring to the hereditary developmental resources handed down from parents to children via biological reproduction and *nurture* as an inclusive term for culture, environment, and everything else not transmitted via biological reproduction. This divide became crucial for studying a new field-defining explanandum: heredity.

In the nineteenth century, historians such as López-Beltrán (1994, 2007) claim, an intellectual shift happened: reference to the adjective *hereditary* (e.g., as in the idea of "hereditary disease") was increasingly replaced by a nominal use. The noun *heredity* became established as the name for a new field-defining phenomenon, something in need of explanation. This amounts to a reification with new ontological commitments, that is, the creation of a new epistemic space, as Müller-Wille and Rheinberger (2007) stress. Most important for this study, it

> also implies a concomitant shift, namely the erosion of a set of very ancient distinctions with respect to similarities between parents and offspring, which the modern notion of heredity systematically cuts across. Distinctions had been made between specific versus individual, paternal versus maternal, ancestral versus parental, normal versus pathological similarities, and even between similarities pertaining to the left and the right halves of the body. Such distinctions gave way to a *generalized notion of heredity* that focused on elementary traits or dispositions independent of the particular life forms they were part of, whether pathological or normal, maternal or paternal, individual or specific. (12–13, emphasis added)

The discussion in chapter 3 indicates that a populational stance came with that shift, a move away from individuals and their intrinsic properties toward the idea of populations, first of individuals and then of the developmental resources hosted by these individuals. Thus, Galton assumed a pool of biologically inherited developmental resources as explanatory for evolution, development, and heredity. What is most important in this chapter is that Galton replaced the older distinctions mentioned by Müller-Wille and Rheinberger with a new generalized distinction: the nature-nurture divide.

The establishment of this contrast was strongly influenced by the anti-Lamarckism in Galton's views about heredity. The anti-Lamarckism has led to a separateness of nature and nurture that influenced the "century of the gene" (Keller 2000) as part of which ideas about genetic factors (equated with nature) were regarded as more important than other developmental resources. To understand the connection to Lamarckism is crucial to understand what exactly is challenging about the nature-nurture distinction.

Galton's Anti-Lamarckian Elements

Galton is also taken to be the person who strengthened the idea of particulate inheritance, also defended by Charles Darwin. It is an idea that took biologically inherited developmental resources as material particles (i.e., substances)—*gemmules* in Darwin's case and *stirps* in Galton's. For Galton, these hereditary units were material and internal to individuals (as for Darwin) and they were (contra Darwin) fixed, that is, unchangeable. They were as ahistorical as the units of the physical and chemical world were believed to be: they were elements—"elements of which personality is composed"—and therefore cannot be changed during individual development. Galton's distinction between stirp versus person (latent versus patent elements) consequently counts as a predecessor of the germ-soma distinction, later introduced by August Weismann, as part of his anti-Lamarckian theory of inheritance.

Galton not only believed that developmental resources are ontologically like physical elements, but also that they are causally like these. He believed that their effects could be treated like the effects of physical forces—in a commensurable, additive manner, with each cause having an effect of its own. Galton conceived the effects of causes not only as separate but as comparable, as being commensurable.

Nevertheless, Galton agreed that one needs nature and nurture to explain, for instance, the development of an individual's body height. This is often ignored. The very book in which Galton introduced the "convenient jingle of words" was a reply to Alphonse de Candolle's critique of Galton's earlier works. Candolle claimed that Galton ignored the influence of culture and environment. Far from it, Galton replied in his *English Men of Science*, and added that he could even measure and compare the effects of both these kinds of causes.[3]

With Galton, developmental resources became divided not only into (1) two different kinds of causes (his elements), but also in a way so that (2) the effects of these kinds of causes can be described quantitatively and additively (with the help of statistics) as a distinct, relative contribution for a specific phenotypic variance of a trait. All that talk about something being in a certain percentage due to nature rather than nurture derives from this second aspect.

Because he introduced this change to a quantitative explanation of population variance, Keller (2010) takes Galton, as others do, as the origin of an opposition of nature versus nurture—the hardening of the divide. I disagree. The decisive point is not the quantitative account that stems from the idea of commensurability but his anti-Lamarckism. First, John Stuart Mill (1858) was the person who introduced in his *System of Logic* the idea that some causes can be compared in the way Galton did. According to Mill, sometimes (e.g., in Newtonian physics) causes are commensurable so that the effects can be apportioned to the separate causes. Consequently, the causal contribution was conceived as being added onto each other in a quantitative manner. This is Mill's principle of the composition of causes (Mill 1858, 3.6). All Galton did was apply the "Newtonian approach to apportioning causal responsibility," as Sober ([1988] 1994, 185) calls that way of dealing with causes, to the realm of explaining behavioral differences; introduce the necessary statistical methods to calculate the magnitudes; and pioneer twin studies as a natural experiment to get the data for the calculations about it (i.e., about heritability in a narrow sense).[4] Second, as Paul and Day (2008) point out, Mill and others were well aware that nature and nurture can act in opposite and therefore separate directions. Paul and Day even give evidence that talk about the *"relative importance* of innate characteristics and institutional arrangements in explaining human difference" (emphasis added) was already present during the first half of the nineteenth century. Galton added only the methodology to quantify the relative importance of nature and nurture for the development of a trait. Thus, as in chapter 3, where Darwin and Galton were treated as centers of gravity for the Darwinian challenge, Galton is "best viewed not as initiating a debate but as intervening in one that, in the terminology of the time, counterpoised 'innate character' to 'institutional arrangements' in explaining human mental and moral difference" (Paul and Day 2008, 222).

This does not mean, however, that there is not something that also conceptually (not just methodologically) divides Galton from others of his time, be it Mill or Darwin. If Galton is the one who created the "mirage of space between nature and nurture," as Keller (2010) calls it, then this is not only because he regarded nature and nurture as separate causes able to act in different directions or because he initiated an effect-apportioning statistical method. It is because he also believed in anti-Lamarckian hard heredity, against most of his fellows and also against his cousin Charles Darwin. Others, such as Wallace and, most famously, Weismann, quickly joined in, to form a coalition against Lamarckian inheritance.

In sections 4.2 and 4.3, I show how commensurability is criticized on the basis of the interactionist consensus. Since, as I also show, anti-Lamarckism still holds, nature as distinct from nurture survives the developmentalist challenge, despite epigenetics and similar interactions of nature and nurture. Before I can go there, the importance of the anti-Lamarckism for the nature-nurture divide needs to be defended in more detail.

The Importance of Anti-Lamarckism for the Nature–Nurture Divide, or How Culture Became Autonomous and Nature Universal

Anti-Lamarckism made it conceivable that culture is autonomous and human nature shared across all human groups. Autonomy of culture and antiracism are the two issues that will help us see the importance of anti-Lamarckism for the nature-nurture divide. After Galton, it was mainly Weismann (1891, 1893) who became known to deny that inheritance of acquired characteristics is possible. He claimed that there is an "all-sufficiency of natural selection," meaning no Lamarckian inheritance is necessary for evolution, whatever case one chooses. Herbert Spencer (1893, 1894) famously disagreed and argued for the necessity of Lamarckian inheritance to explain the evolution of musical sense, and mind and higher cognition in general.

The Developmentalist Challenge

No genius (e.g., Mozart, an example used in the debate) without Lamarck (i.e., Lamarckian inheritance): that was Spencer's position. Weismann, by contrast, believed that only if Lamarckian inheritance is replaced with the concept of cultural inheritance, conceived as running autonomously and parallel to biological inheritance, is one able to see that cultural differences (in space) and cultural change (differences over time) happen autonomously, decoupled from biological change. As a result, culture became conceptualized as autonomous and nature as universal.

• *Culture as autonomous.* Weismann's perspective came to full bloom in Alfred L. Kroeber's cultural determinism. Kroeber (1917), an anthropologist, depicted the autonomy of culture in a famous figure (see figure 4.1). The figure clearly shows how culture takes off. There are then for the anti-Lamarckian, two separate evolutionary processes: biological evolution and cultural evolution. For the Lamarckian, however, culture was coupled with and reducible to nature since culture slowly but steadily and repeatedly becomes nature, habit becomes instinct, acquired becomes innate—all via the biological inheritance of acquired characteristics. In the Lamarckian picture, the two kinds of evolution might be conceptually separate, but they are causally correlated: if one changes, the other does too. The takeoff in the figure from Kroeber would be impossible if Lamarckian inheritance reigned. In addition, the takeoff of cultural evolution amounts to an underdetermination of culture by nature because of a one-to-many relationship:[5]

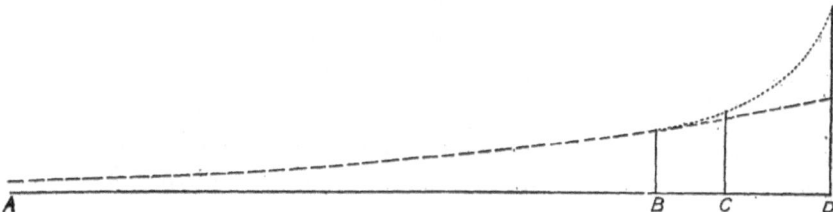

Figure 4.1
Culture taking off. The horizontal axis represents time and physical persistence, which is taken to not evolve at all. The vertical axis represents change (accumulation or increase in complexity even). Cultural evolution (the dotted line) is autonomously changing in relation to biological evolution (dashed line). B is the first animal using culture, C the beginning of our species, D the end of the nineteenth century. (Reprinted from Alfred L. Kroeber, "The Superorganic," *American Anthropologist* 19, no. 2 (1917): 163–213, 211. Public domain.)

one human nature is historically or synchronically connected with many human cultures. Thus, the "one human nature" cannot explain the differences between the "many human cultures" (synchronically or diachronically). If nature is not making the differences we see in cultures and therefore cannot explain culture, then only culture can explain culture.

• *Nature as universal.* As long as nature and culture were coupled via Lamarckian inheritance, group-related cultural differences could be taken as indicators for biological differences. This is the reasoning that stands behind racist leanings in Herbert Spencer's philosophy. Spencer's belief in the inheritance of acquired characteristics was used not only to explain the evolution of mental abilities (e.g., musical ability) but also to claim that cultural differences correlate with racial differences. Given the Lamarckian coupling of nature and culture, biological differences could be inferred from cultural differences (and vice versa). This is why for Spencer, the psychic unity of mankind was impossible: cultural differences between people amount to biological differences, by necessity of Lamarckian inheritance. That way, racism (understood to be about group-related biological differences) found a stable justification in Lamarckism. On the basis of a Weismannian point of view, however, one could not infer racial differences from cultural differences since the two were regarded as independent, decoupled from the very first moment when the first animal managed to learn socially from another one, that is, from the very birth—if there was any concrete such date—of culture via nature, represented as point B in Kroeber's famous figure (figure 4.1). Since a non-Lamarckian frame does not allow to infer racial differences from cultural differences, a non-Lamarckian such as Kroeber saw no evidence to not believe in the so-called psychic unity of mankind, that is, a human nature in the sense of one humankind being psychologically on a par, with all human groups sharing the same psychic abilities.[6] This is how non-Lamarckian thinking strengthened the idea of one universal nature shared by all human groups (though not necessarily by every individual).

Culture as the Fast Track of Evolution

Decoupling biological and cultural evolution was certainly not providing concluding evidence for the unity of humankind, but it made racism less likely. It was, first, blocking a specific, a Lamarckian argumentation pattern for racism that had been used until then. It was, second, providing some new evidence against racism: since culture was—as a consequence—conceptualized as able to take the lead in evolutionary problem solving, evolution could be understood to mainly happen via culture, preventing biological differentiation of the one humankind. I discussed the first above and discuss the second briefly now. Weismann and Kroeber's perspective allows the argument that the evolution of biological differences (matching the evolution of cultural differences) is unlikely since if nature and culture are two separate decoupled evolutionary processes, then it is likely that the takeoff one sees in figure 4.1 exists. This is so since culture is a better, more efficient process of adaptation since it can react more quickly. This in turn makes it likely that biological evolution has come to a virtual halt. In other words, once culture is available, problems can get solved via culture—the fast track of evolution. As a result, biological evolution can slow down tremendously, virtually (i.e., relatively to the tremendous change in culture) not changing anymore.

To sum up, with Galton, Weismann, and Kroeber, nature and culture became separated in a very specific sense because of the anti-Lamarckism involved. As a consequence, culture was taken to be autonomous and nature universal (i.e., applied to all groups of humans).

4.2 Ignoring Interactions

Separating nature and culture (and also environment) in the way described has nonetheless been attacked because using it leads people to ignore the interactions of nature, culture, and environment at the developmental, intergenerational, and evolutionary levels. In this section, I introduce four main bodies of literature that are well known for a critical stance along these lines: anthropological literature, dialectical biology, developmental systems theory, and discussions on epigenetics.

Anthropological Literature on Alternative Ontologies

The anthropological literature (e.g., Strathern 1992; Viveiros de Castro 1998; Ingold 2000; Descola 2005) has stressed that other cultures do not have a vision of humans as being composed of two categorically separate levels or principles, nature and culture, with the latter on top of or added to the former, culture being at best a result of the more fundamental principle of nature. Sahlins (2008) even speaks of the "Western illusion of human nature" that is built on the age-old contrast among nature, culture, and environment. His critique is directed at the intellectual poverty and biasedness of a view that regards humans as beasts due to nature. To counter it, he stresses the importance of culture for evolution, as well as for development. For him, culture is not on top of nature, as in Kroeber's separationist picture, since human nature is the evolutionary result of a complex entanglement: culture and nature interacting constantly and in non-Lamarckian manners at the developmental and evolutionary levels. I will come back to this aspect when I discuss coevolution of nature and culture as part of the interactionist consensus.

Dialectical Biology, Developmental Systems Theory, and Epigenetics

From within discussions about development and evolution, Lewontin (1983; see also Levins and Lewontin 1985) has become a spokesperson for so-called dialectical biology, which regards the organism and the environment in a dialectical relationship of coevolution. Rather than taking the organism as a passive object, Lewontin's dialectical approach takes the organism as subject, as actively shaping its niche and thereby also as directing its own development and evolution. Developmental systems theorists like Oyama (1985) defend a similar position, but they are more critical of the nature-nurture divide. It is not enough, they claim, to conceptualize nature and nurture as interacting and to take into account that there is a dialectic relationship between organisms and environments and a lot of developmental plasticity, as West-Eberhard (2003) has stressed. The divide itself has to go. They suggest regarding "nature" either as the holistic overall typical result of development or as an overall set of all developmental resources. A huge developmental system is all there is, with no need for a nature-nurture divide. They believe in "one process, indivisible," as Schaffner (1998) wrote.[7]

Recently the perspective of developmental systems theory has been complemented by discussions about epigenetics, here understood as a part of molecular biology, which studies the effects of social or environmental factors on molecular processes, mainly gene expression. If these effects are transmitted to the next generation of individuals, then there is epigenetic inheritance. Epigenetic inheritance can then be defined as the biological transmission of genome expression patterns that result from epigenetic processes—processes that regulate gene action without changing the DNA. Examples are mainly chromatin modifications transmitted over generations of organisms. Jablonka and Lamb (1995, 2005) were the first to campaign for the importance of epigenetic inheritance in that narrow sense. Currently, a lot of research in social as well as natural sciences is concerned with epigenetics.[8]

What is their importance for the nature-nurture divide? Epigenetic effects diminish the general importance of genes for explaining development and the resulting behaviors or phenotypic traits. Epigenetics showcases how much interaction there is between nature (understood in that context as developmental resources traveling via biological reproduction), culture (developmental resources traveling via social learning), and environmental developmental resources (those that do not depend on any social learning for their intergenerational presence, e.g., the sun).

Epigenetic interaction happens during development, and some of the effects of the interactions are transmitted by biological reproduction, that is, the biological inheritance from one organism to the next. The challenge it provides is not that genetic factors are changed since only the gene expression is changed (i.e., how much and when a gene is expressed). The challenge is, rather, that the interactions between nature and nurture are so strong that there is dynamically a coupling of nature and nurture, similar to the Lamarckian coupling already described: if nurture changes (and this has epigenetic effects), then nature (what is inherited biologically) changes too, since the expression patterns (not just the genes themselves) are biologically inherited, even if only for a few generations. Nature becomes malleable again by culture. Consequently, racial differences (i.e., biological differences between groups) resulting from cultural differences become an important topic again, as discussed, for instance, in Kuzawa and Sweet (2009).

To sum up, the perspectives of these kinds of critiques—the anthropological, the dialectical, the developmentalist, and the epigenetic critiques—are important in order to illustrate the poverty of stances that reduce everything to biologically inherited factors (nature), not to speak of those still believing in genes as master molecules. Genetic determinism, genetic reductionism, and gene centrism (and any essentialism relying on it) are as dead as they can be because of the interactionist consensus, the ultimate core of the developmentalist challenge. I will reconstruct the interactionist consensus in the following section in a systematic manner. This then allows us to see in a more precise manner what is controversial with respect to the nature-nurture divide.

4.3 The Interactionist Consensus

The interactionist consensus states that nature and nurture interact at all levels (the developmental, short-time epigenetic, and evolutionary level) and in ways that make it dubious to consider a trait as being due to nature alone. I take this interactionist consensus to embrace five claims that need to be distinguished clearly: genetic inertness, incommensurability, quantifiable difference making only, epigenetic interaction at the intergenerational level, and interaction at the evolutionary level.

I take the simple example from the previous chapter, human body height, as the phenomenon to be explained. The example is simple in the sense that the trait itself is concrete and easy to measure. That way, I can stay fixed on the problem of the interaction of nature and nurture as kinds of causes (i.e., as sets of developmental resources) without other issues getting in the way, for example, abstraction and operationalization of the phenomena explained, issues that have to wait until part II.

Figure 4.2, a so-called norm of reaction graph, displays how one can measure the causal influence of nature and nurture on a trait by plotting norms of reactions.[9] Such a norm of reaction elicits causal information from a rather simple setting, incorporating the one-gene-to-one-phenotype picture of the early time of the "century of the gene," to use Keller's (2000) phrase once again. Norms of reaction incorporate a causal structure: a representation of two candidate factors (factors that are likely to be causally relevant given prior knowledge), one genetic and one environmental.[10] All other causal factors are either ignored, because they are unknown, or held fixed at a certain value or randomized because they are not of interest. The goal is to find out whether the two selected causal factors make a difference, and, if so, which or even how much difference they make for a respective explanandum such as body height. The developmental processes (or mechanisms, if you like) leading to the trait are completely black-boxed.

To claim that nature and nurture developmentally always interact so that it is unreasonable to call a trait such as body height due to nature

The Developmentalist Challenge

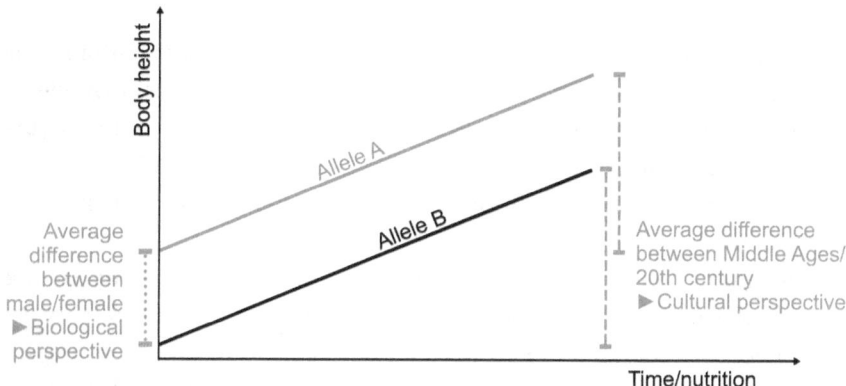

Figure 4.2
Hypothetical norm of reaction for human body height. The lines for alleles A and B in this hypothetical graph (respectively, gray and black) are visualizing data about body heights of two groups of humans, one with allele A and the other with allele B. The lines represent the so-called norms of how the two groups at issue—in this case, human male and female—behave regarding the phenomenon at issue (human body height, measured along the vertical axis), given variation in an environmental variable (measured along the horizontal axis). The environmental variable is assumed to refer to nutrition as having changed over time from the fifteenth to the twentieth century. Since the lines are here assumed to be parallel, there are two differences regarding the phenomenon that stay the same: the averaged difference between male and female height (vertical dotted line on the left) and the average difference between the Middle Ages and the twentieth century, all humans taken together (vertical dashed lines on the right of the same length). The first difference is interesting from a biological perspective, the second from a cultural perspective. (The assumptions visualized here are taken from argumentation presented by Fukuyama 2002, 129–133. Illustration: Vera Brüggemann, Bielefeld.)

amounts to one or a set of the following three points: genetic inertness, incommensurability, and quantifiable difference making.

Genetic Inertness

First, any trait will need some environmental factors to develop. Genes are inert; they cannot do anything on their own. For instance, without any nutrition, there will not be a body and, hence, no body height to measure. In that rather trivial sense, nature and nurture always interact causally.

Incommensurability

More important, it is now a consensus that the causal contributions that genes and environment make to a phenotypic trait are not comparable in any quantitative manner. They do not come in the same "common currency" and are not "commensurable," to use Sober's ([1988] 1994, 195–198) terms. Consequently, one cannot say that one is more important than the other.

Lewontin ([1974] 2006, 2) compared the incommensurability of the causal relevance of nature and nurture to the incommensurability of the causal contribution of two people building a wall, the one bringing the bricks and the other the mortar. Imagine the following: Suzy brings the bricks and Billy the mortar. We can measure for each how much higher the wall gets in a certain amount of time (say, a day), depending on the speed of one of them, holding the contribution of the other fixed. So given that Billy moves at a constant, abundant speed, one can measure how much Suzy contributes to the difference in the wall's height, when she gets normal nutrition and compare it to how much the wall changes when she gets almost nothing to eat for the day. We can then set Suzy constant and vary the nutrition on Billy to see how much difference it makes when his nutrition changes. In both cases, we will see whether, which, and also how much difference it makes for the wall to intervene on Suzy or Billy. Yet, and that is the important point, one cannot get an answer for the following question: How much of the wall is due to Suzy and how much to Billy? This is a meaningless question, a pseudo-question without any possibilities for an answer; the contributions of Suzy and Billy cannot be compared quantitatively since they contribute in qualitatively different ways to the wall. One brings bricks, the other mortar. One cannot look at the wall and say, "Ah, 60 percent of the wall is due to Suzy." Stones and mortar do not and cannot come in a common currency. There is no way to bring the two causal factors, Suzy and Billy, into direct, quantitative, and commensurable comparison.

The same holds for nature-nurture issues—that is, for "building" people. In developmental processes, different kinds of causal factors interact, each kind coming in its own currency, with no conversion table. We therefore cannot apportion the causal influence of two necessary factors for development (e.g., a gene and a nutritional factor) in an additive manner (just

The Developmentalist Challenge 73

quantitatively adding them up), asking how much of body height nature causes in comparison to nurture.

It is important to keep incommensurability distinct from genetic inertness: that there is no commensurability is not only because there simply will not be anything to measure in the absence of nature or nurture. Even if there is a body, and one can thus measure something about nature and nurture's causal importance, one cannot measure what some would like to know, namely, how much each factor contributed to the trait in and of itself. By contrast, imagine a wall that needs only two kinds of commensurable ingredients (e.g., a pure concrete wall that needs two kinds of concrete, both measured in kilograms, rather than a wall built out of stones and mortar). Imagine that there would be inertness of the one causal factor (concrete 1) without the other (concrete 2). In such a case, there would still be commensurability since both come in the same currency: kilograms of concrete. In fact, one can do such an apportioning sometimes when one cites physical forces, as when one explains what causes the speed of a physical object on an inclined plane (an example most will remember from physics class). That one cannot do such a quantitative causal apportioning for the development of living entities is a point that John Stuart Mill already acknowledged when he introduced in his *System of Logic* (1858) the apportioning way of causal thinking. Galton was the one who thought that heritability measures are a methodological solution for apportioning causal relevance even for living entities. Yet the problem is that heritability measures are a technical solution for statistical correlation rather than a philosophical one for causal relevance. In addition, it has—even as a technical solution—its problems, which need not be of concern here.[11]

Quantifiable Difference Making

As the example of Suzy and Billy bringing bricks at different speed should illustrate, all one can do experimentally is to measure how much quantifiable difference nature and nurture each can make on its own to the trait. That causes are difference makers is an important aspect of the concept of causation. I am using the concept of difference making in a manner as neutral as possible, since it can be spelled out in different ways. Yet it is no coincidence that the language of difference making goes back to John Stuart Mill, who first noted the incommensurability of causes in the above explicated sense and already provided a solution with the idea of understanding causation as difference making.[12] I argue that difference making (but not causal contribution) can be measured in one currency, either in units of the effect variable (e.g., height of the wall, height of a body) or at least in terms of increased risk (i.e., in terms of probability of the trait occurring increased by the respective causal factor).[13]

Thus, given that the causes (and their causal contribution) are incommensurable in the described sense, one can only measure the effect (i.e., the difference the cause makes) in the currency of the one and the same kind of effect. The currency of effect in the toy example of the wall is the amount of wall per time unit; the currency of effect in developmental process is an amount of a developmental property such as body height. With that, incommensurability is not overcome, but causal relevance and some pattern of (in-)dependence between nature and nurture are rescued, despite incommensurability.

Causal relevance and patterns of (in-)dependence In the hypothetical norm of reaction in figure 4.2, one clearly sees that both nature and nurture are making a difference and are therefore causally relevant for the trait: changing the gene makes a difference to height, and changing the environment does too. Test it: stay with one environment, and change from allele A to allele B, and the body height differs; stay with one allele (e.g., allele A) and change the environment, and the body height differs again. Nature and nurture both make a difference. That they both make a difference can be measured simply because an effect on the trait can be observed when one changes one thing at a time.

Which difference (i.e., how much exactly) the causal factors make can also be measured in our hypothetical example of measuring human body height. If one imagines again the hypothesized difference the gene makes for body height and then also varies the environment, the difference the gene makes is always the same (let's assume it as 10 cm difference). The dotted vertical line at the left side of the picture stays the same if one were to move it to the right, to follow the change in environmental variable. In such a case, the difference the gene makes is independent of the environmental factor. Now, if one imagines the difference nutrition makes, then that difference is equally independent of what happens with the gene since the dashed line (the vertical line at the right side of the picture) is the same (let's assume it makes a 30 cm difference), independent of whether it results from differences with respect to allele A or allele B. In such cases, the effects (of the gene or the environment) are stable across the measured environments (in the case of the gene's difference making) or across the measured alleles (in the case of the environment's stable difference making). If a norm of reaction has such a structure, then, in the language of quantitative studies, there is a genetic main effect and an environmental main effect.[14]

This can then, however, nonetheless be the basis of a sleight of hand that, for example, Fukuyama (2002) performs when he discusses the example of human body height. He admits that human body height increased over the centuries due to the marvelous better nutrition humans have today compared to our Middle Ages ancestors in the fifteenth century. Fukuyama thus joins the chorus of the interactionist consensus: yes, body height, sure, is due to nature *and* nurture. A few lines later, however, he states that nonetheless "the average male-female differences are the products of heredity, and thus nature" (130–133). He is talking about "products of heredity" and therefore using causal language. Is he contradicting himself? No; he is simply playing the rhetorical sleight of hand that Gannett (1999) and Keller (2010) already criticized, a sleight of hand of implicitly *reconstituting the phenomenon* (changing the explanandum), from explaining a trait (a concrete property such as body height) to explaining differences regarding a trait (the average male-female difference in height, represented by the vertical dotted line, on the left side in figure 4.2). What Fukuyama says is, yes, certainly, changing the environment makes a difference to height, the trait, but it does not make a difference to the difference between male and female average heights. As noted, the hypothetical average difference between average male body height and average female body height that is made by the gene is depicted by the dotted vertical line at the left side of figure 4.2. It is interesting for geneticists simply by disciplinary affiliation. Geneticists want to know whether a gene makes a difference. That is their job. They consequently watch out for the difference the gene makes a difference to and ignore the rest, explanans-wise and explanandum-wise.

Others have a different job and might reply: "Hey, that is hereditarianism and wrong since we agreed that nutrition makes a difference, didn't we?" The biologist answers, "Well, we did, but making a difference to what? It makes a difference to height; it does not make a difference to the difference between males and females," pointing to the dotted line at the left side in figure 4.2, which stays the same whatever one does with the depicted environmental variable. "Ah, I see, right, but wait a minute," the cultural anthropologist goes on. "The environment also makes an exclusive difference, but to a different difference: it makes a difference to what is represented by the dashed vertical line at the right side of the picture, namely, the difference in average height of humans (all together) between the Middle Ages and today. This is the difference I, as a historian and cultural anthropologist, am interested in. It is a difference to which your genetic factors are not making a difference." At the end, there are in this example two perspectives, a biological and a cultural one, with different differences as respective explanandum. Each points to a respective pattern of developmental (in-)dependence that is best captured by the parallel lines in a norm of reaction. Understanding an item "of culture," that is, one that varies in time or place (e.g., concrete languages), is explained by "culture" alone, if differences in biologically inherited factors can safely be ignored for the difference between the two endpoints in the historical or spatial comparison. Those who simply repeat all the time that nature and nurture are always interacting will be unable to see such interesting patterns of (in-)dependence, as Kitcher (2001a) already stressed, and convincingly so.

Only if the two lines were to collapse into one would the gene fail to make a difference to the trait, and only if the two lines were completely horizontal would the environmental variable fail to make a difference to the trait. Only the latter, a flat norm of reaction, a very rare and specific pattern of independence from the environmental factor, would be evidence for genetic determinism (but not proof, since there are problems of extrapolation to unmeasured cases that I discuss below). A standard example of a trait with a quite flat norm of reaction is phenylketonuria (PKU), a rare disease that is regarded as genetic since certain mutations to a specific gene (the gene for PAH, short for the gene that codes for the enzyme phenylalanine hydroxylase) cause the disease. The "reaction" (effect of the gene interacting with environmental factors) is always the same—there is no range or norm of reaction, only a flat line—except for one case: when a strict diet that prevents any intake of phenylalanine is kept.[15]

Usually the situation is not even close to being so "straight," even without extrapolation, to which I will turn shortly. The hypothetical case of human body height is already an idealized case. In reality, norms of reaction often look pretty wild. The difference the one factor makes (e.g., an

The Developmentalist Challenge   77

increase of value from allele A to allele B of 10 cm in height, represented by the dotted line on the left of figure 4.2), would then not stay the same if one moved from one environment to the next (from left to right in figure 4.2). Such cases are also called nonlinear.[16] The wild norm of reaction in figure 4.3, for instance, has become well known in the literature.

Nevertheless, even such actually measured norms of reactions can sometimes reveal interesting patterns of dependence between nature and nurture.[17] For instance, when the lines cross, the value of the environmental variable makes a difference to whether (and how much) one allele increases or decreases the trait (compared to the other variant of the gene). The direction of effect of one factor is then dependent on the value of the environmental factor. If there is such a dependence of the kind of effect of the genetic variable on a specific value of the environmental variable (and vice

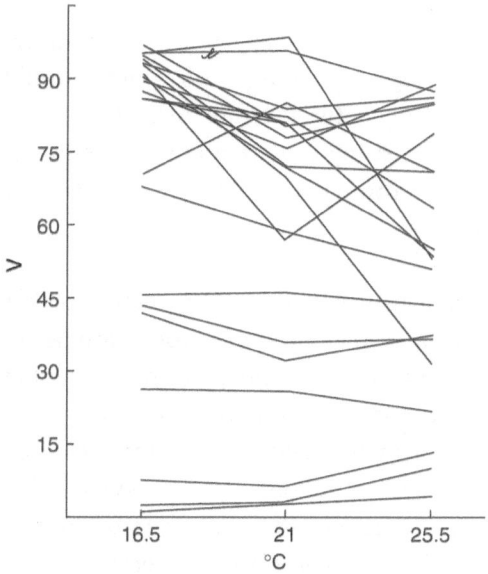

Figure 4.3
A nonhypothetical norm of reaction. The graph depicts viability of different types of *Drosophila pseudoobscura*. It shows that viability V (measured at the vertical axis) depends on genetic factors as well as temperature (measured on the horizontal axis) and does so in a nonlinear way. (Data are based on Dobzhansky and Spassky 1944. Reprinted from Richard C. Lewontin, "The Analysis of Variance and the Analysis of Causes," *International Journal of Epidemiology* 35, no. 3, 2006 [1974]: 520–525, 524. With permission from the author. © Richard Lewontin.)

versa), then this is usually called *gene-environment interaction*, which I call *narrow interaction* since it is an interaction different from the trivial concept of interaction already mentioned.

A recent case for such a narrow kind of interaction has stirred some debate. Caspi et al. (2002), and later studies, established some evidence of a narrow interaction with respect to the so-called MAOA gene, taken to be causally involved in the ontogeny of aggressive behavior.[18] What Caspi et al. (2002) showed is that the difference the respective MAOA allele makes, in terms of the probability of aggressive behavior (whether the probability is increasing or decreasing), depends on whether the nurture-variable takes on a specific value. They found that abusive environments increase the probability of aggressive behavior, whereas nonabusive environments decrease it, but only for a specific allele since the situation is upside down for the other allele. The respective norm-of-reaction figure would show a cross between the two lines of the alleles of the MAOA gene that have been tested. In such a case, which difference a specific allele makes (increasing the probability of aggression or decreasing it) depends on which cultural factor is present. In case of such a narrow nature-nurture interaction, there are opposed effects of nature (i.e., increasing or decreasing probability of aggression), depending on inputs of culture, and equally opposed effects of nurture, depending on input of nature.

Extrapolation A final and important issue regarding these matters is that one cannot extrapolate from the measured environments or genotypes to unmeasured or unconceived ones, and this holds independent of whether there was an interaction in the narrow sense measured or not. Even if the lines have been parallel, horizontal, or what have you, for the measured values of the cause variables, the generalizations reached always hold only for what has been measured.[19] There are no strict generalizations in developmental biology since they always involve a lot of ceteris paribus conditions.

This is not just of abstract theoretical importance. It has political importance too. If the problem of extrapolation is taken seriously, any "it-cannot-be-otherwise" conclusion is misleading and politically dangerous, since whether a gene or environment makes a difference can always turn out to be dependent on the choice of values tested for the two cause variables. Whether and how much difference genes or environments make is thus dependent on a choice of possible worlds considered: it matters which possible world one chooses, as already Sober ([1988] 1994) made vivid.[20]

For instance, from all science has learned about body height, it is likely that if one checks back five hundred years from now, the body height of humans will not have increased much more. There are limits of growth for humans, regardless of how good nutrition becomes. Thus, if one measured only the influence of the environment in two or three slightly differing nutritional heavens—possible worlds utterly and shamefully far away for shockingly many human beings—then the nutrition would have turned out to not make a difference (or close to none) since the norm of reaction in these worlds would be flat (or close to flat). The scientist doing these studies could conclude that body height is genetically determined, given the evidence available. But it is genetically determined only in these chosen worlds. If the scientist had measured the historical world from the fifteenth to the twentieth century or the worlds of those less well off, then body height would have turned out to be not genetically determined. The dependence on our choice of a possible world will be important in chapter 9 since it clearly has a political dimension. We *make human nature* by, first, choosing a narrow set of possible worlds that suits the interests of those seeking an explanation (often those in power), through which we, second, create (or stabilize) that world.

To summarize, there are four important aspects of interaction at the developmental level: first, nature and nurture interact at the developmental level in a trivial sense because of genetic inertness; second, there are decisive limits of deriving knowledge about a specific causal factor, in terms of both comparing the effects (incommensurability) and in terms of extrapolation; third, there are nonetheless specific patterns of dependence observable, within the limits set by incommensurability and problems of extrapolation.

Interaction at the Intergenerational Level
There is also interaction of nature and nurture at the intergenerational level. Philosophical interpretations of epigenetic inheritance (e.g., Lock 2013; Meloni 2016) have stressed that the causal influence of nurture on gene expression is establishing an entanglement of nature and nurture that amounts to more than the broad and narrow interaction already mentioned. Epigenetics shows that "life events and environmental exposures

are literally embodied," as Lock (2013, 292) writes. This gives further evidence to the idea that biology is "local biology" rather than universal. This entails a critique of the tendency to normalize bodies into "universal bodies." Even more important, at least for the context of this study, these local and therefore encultured bodies have an intergenerational effect via epigenetic inheritance. Jablonka (2016, 55) writes, "At the conceptual level, acknowledging the existence of epigenetic inheritance renders the traditional nature-nurture dichotomy obsolete, because it means that heredity ('nature') can be developmentally constructed ('nurtured')."

It is, in this study, not so important whether it is the language of embodiment or the language of construction that is used for the epigenetic entanglement claim. What is important is that both perspectives point to an effect of nurture (or culture) on nature that allegedly amounts to more than the broad and narrow interaction already introduced. In the following, I argue that this is misleading. As part of this, I argue that the kind of embodiment that Lock and Jablonka refer to also happens in broad and narrow interaction.

First, with Lamarckian heredity off the table and genes at the center, the belief in the twentieth century was that the life events of the parents are irrelevant for what the next generation biologically inherits. Since Weismann, not just biological inheritance has been decoupled from cultural inheritance, but also the germ cell line from the somatic cell line. Life events no more loaded the dice of the biological lottery of sexual reproduction for the next generation. Given epigenetics, parents and everybody else are again, in part at least, responsible for the developmental resources handed down to children by biological reproduction, since some of these biologically inherited resources, the epigenetic ones, are laden with the effects of the parent generation's doings and sufferings, be it via nutrition, addiction, abuse, or injustice. The intergenerational perpetuation of racial injustice via epigenetics is one of the cases currently discussed with that issue in mind by Kuzawa and Sweet (2009) and others. Epigenetics shows us that racism produces race: it leads to embodied racial discrimination, as Meloni (2015) notes, via epigenetic inheritance across generations, ultimately going back to epigenetic processes in one generation.

To acknowledge the epigenetic, intergenerational interaction between nature and culture is important. Yet it does not amount to a kind of embodiment or construction that is stronger than what is entailed in the broad and narrow interaction discussed in the previous section. The molecular mechanisms of how abusive environments are influencing the MAOA gene that Caspi et al. (2002) studied are still being researched. Yet it is likely that it involves something similar to what is now often called *parental effect*, which are epigenetic. A well-known example for a parental effect is the following. Whether a rat mother is licking and grooming her pups affects the expression of a certain genetic factor involved in stress reactivity, which affects the later behavior of the pups. Such a parental effect on gene expression can even result in the reoccurrence of the respective licking and grooming behavior; once the former pup becomes a mother, it repeats the behavior of its mother (Meaney 2001). So it looks like inheritance, right? It's not, for two reasons.

First, just like abusive behavior of human parents (as described by Caspi et al. 2002) has an effect on whether the respective children will become aggressive, neglecting behavior of rats has an effect on the bodies of their offspring. Thus, parental effects in and of themselves simply involve a kind of embodiment. That embodiment, though, is not specific to epigenetics since any interaction involves that kind of embodiment, for example, the embodiment of aggressive behavior of parents (or other caretakers) in the next generation's aggressive behavior. Every effect of nurture on bodies is a case of embodiment (or construction). If a certain kind of nutrition interacts with genetic factors (in the broad sense of interaction) to cause height in plants, one has a case that one could call embodied nutrition. The same holds for all other parental effects. What is most spectacular about epigenetics is not that there are embodied parental effects, but that some of these embodiments can be transmitted by biological reproduction to the still next generation. It can persist for a few generations at least, even if the environmental cause disappears.

Second, although parental effects influence gene expression and although they can be inherited, they are not qua being parental effects a case of inheritance, as Merlin (2017) argues. What parental effects qua parental effects are doing is introducing novelty into the next generation's somatic line. This novelty can be adaptive, but it does not have to be. The novelty can be epigenetically inherited to the still next generation, but it does not have to. As Fish et al. (2004, 167) write in their scientific study on the rat case, parental effects "may have evolved to program advantages in the environments that the offspring will likely face as adults." It follows that parental effects are (if at all) a mechanism of adaptation, not in and of itself a mechanism of inheritance.[21] The moment the novelty gets introduced, the process is one of nature-nurture interaction; the moment the effect gets inherited to the still next generation (if it does), it is simply a normal case of epigenetic inheritance, an inheritance via the germ line to the next body.

The behavioral inheritance we see in the case of licking behavior in rats is clearly based on social interaction, but it is in and of itself either no case of biological germ line inheritance (but a behavioral reaction that might even be classified as social learning) or simply a case of epigenetic inheritance (if there are effects in the germ line that are transmitted by biological reproduction to the still next generation), independent of where the change in gene expression comes from. That behavioral interactions can have an effect at the molecular level (i.e., on the expression of genes in the somatic line of an organism) is important, but it does not show that the distinction between biological and cultural inheritance collapses.

I conclude from all this that since epigenetic inheritance in the narrow sense clearly happens via biological reproduction and since embodiment also happens in normal developmental interaction, neither epigenetic inheritance nor parental effects provide any additional challenge to entanglements at the developmental level. Nonetheless, with epigenetic inheritance, biological inheritance is broadened and becomes soft again. To a limited degree, biological inheritance is influenced by culture via epigenetic inheritance, which sets constraints on the developmental resources available for the future generation, constraints that last a few generations.

In chapters 5 and 7, I show in more detail how to distinguish channels of inheritance, in particular biological inheritance and cultural inheritance, despite epigenetic interaction and inheritance. The distinction between channels of inheritance is also important to understand the interaction of nature and nurture (and, in particular, culture) at the evolutionary level.

Interaction at the Evolutionary Level

Culture and nature interact not only developmentally and across a few generations, but also evolutionarily, despite the autonomy of culture. Cultural inheritance is as evolutionary as biological inheritance. Evolution is therefore not just based on biological inheritance; it relies on multiple channels of inheritance interacting at the developmental, short-time intergenerational, and evolutionary levels. Since culture is a result of the actions of organisms and since biologists consider actions of organisms as part of development, I consider culture as one of the ways in which development and evolution interact at the evolutionary level, in line with Lewontin's dialectical picture. Evolutionary explanations therefore have to take development (and the resulting downstream effects due to cultural inheritance or due to ecological persistence of constructed niches) into account to make sense of the evolution of a trait. Claims defending this now often run under the label "evo-devo."[22]

I will not repeat the standard arguments against gene centrism—the belief that only genes are inherited in the evolutionarily relevant sense. Such a monism is not defendable. Even Dawkins (1976, 1982), known as one of the most radical gene centrists, admitted that culture and ecological niches play an important evolutionary role.[23] What will be discussed is, rather, the holism of developmental systems theory since it denied that it makes sense to talk about a distinct cultural inheritance channel.

A trait is (in the evolutionary sense) evolvable if there is evolutionary heritability of the trait, that is, if parents exhibiting the trait tend to produce offspring who exhibit that trait. Body height, for example, is evolutionarily heritable if parents who are taller than average tend to have offspring who are taller than average. (If one were to plot parental height and offspring height on a grid, the line would have to have a positive slope, increasing from left to right, to be evolvable). Yet even if body height is heritable in that sense, such an evolutionary heritability can still result from cultural inheritance (social learning) and biological inheritance interacting, namely, as difference makers acting every generation over and over again.[24] That parents and children are similar (tall parents tend to have tall children, for example) is a result of developmental resources that travel different channels of transmission. It is a result of similar nutrition (transmitted by cultural inheritance) in combination with similar genetic and epigenetic factors (transmitted by biological inheritance), and similar environmental influences. This is so since cultural, social, and environmental influences also act across generations and can thus stay the same over a long time.

Therefore, these factors can coevolve, a process where cultural inheritance of developmental resources by social learning influences how human nature evolves. Cultural inheritance and biological inheritance do not just run in parallel and cause the similarity of organisms over time; they can also interact phylogenetically. Culture can in such a case even take the lead and direct the selection of a trait.[25]

The evolution of lactose tolerance counts as a paradigmatic example for coevolution of a trait, one that evolved thanks to nature and culture interacting at the evolutionary level.[26] The story behind this example is the following. In some areas, people have heavily relied on dairy farming. As a result of some favorable mutations, some of them had genes that allowed them to digest cow milk even as adults, which fosters in turn dairy farming, which fosters the selection of the respective genes for milk digestion, and so on. This is coevolution, where one has nature via culture and culture via nature, not only ontogenetically but also phylogenetically, even though there are no "genes for" dairy farming and no Lamarckian inheritance, maybe not even epigenetic inheritance.

An important philosophical consequence that can be derived from coevolution is that it revises the dualistic picture about the evolutionary relationship between biological and cultural inheritance. Kroeber's two decoupled processes come together in a new way that explains the evolution of organisms, which is a different explanandum than the evolution of culture. Thus, without contradicting Kroeber's cultural determinism (culture evolves due to cultural inheritance), nature and culture (as separate channels of inheritance) can interact at the evolutionary level, an interaction Kroeber ignored but that is compatible with his picture. Culture can take off, but it can also drag nature along with it. The two inheritance processes can act in accordance since all that an evolutionary sense of heritability (higher similarity between parents and offspring than between offspring and the overall population) requires is that identities are vertical, running in families, (i.e., they are stable across generations).

There is nonetheless an important issue left: if one conceives nature and culture as evolutionary interacting in the way just described, why should one still consider them as separate channels of inheritance? They seem to fail to be autonomous enough, as the argument of Griffiths and Gray (1994, 2001) could be summarized. "So-called channels," Griffiths and Gray (2001, 196) claim, are "not generally independent of one another."[27]

Griffiths and Gray argued for the inclusion of culture as part of evolutionary explanations in a holistic fashion, claiming that the interactions are so strong so that the nature-nurture divide is no longer applicable. They

The Developmentalist Challenge

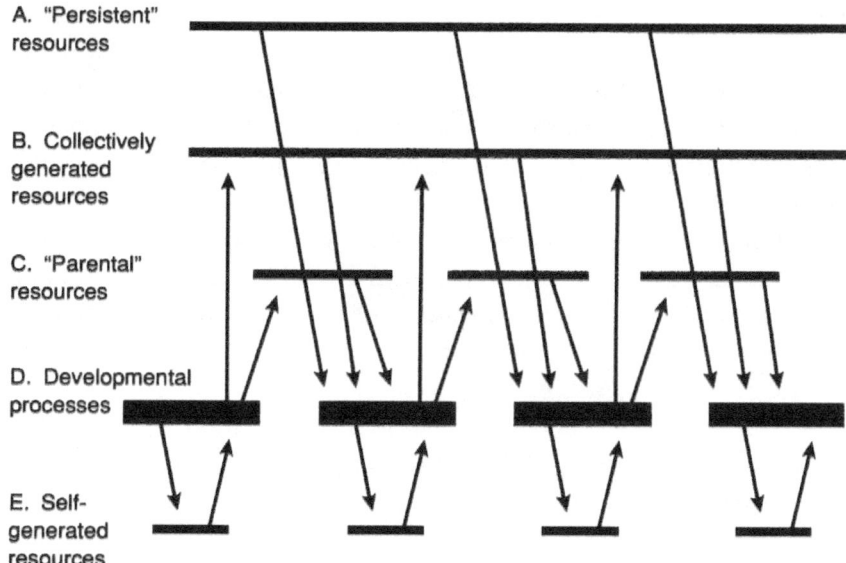

Figure 4.4
Partitioning developmental resources. Developmental systems theorists have partitioned developmental resources in a manner that does not distinguish between a biological and a cultural channel of inheritance. Persistent resources are those that are not transmitted at all; they simply persist. Parental resources are those inherited from parents. The original figure caption states that "the effects of temporal order of interaction have been overlooked. The broad categories of resources are not intended to be exhaustive, and are made largely for convenience of exposition." This will be important in later chapters. (Reprinted from Paul E. Griffiths and Russell D. Gray, "Developmental Systems and Evolutionary Explanation," *Journal of Philosophy* 91, no. 6 [1994]: 277–304, 285. With permission from the *Journal of Philosophy*.)

consequently partition different kinds of developmental resources without using the nature-nurture (or nature-culture) divide, for instance, as depicted in figure 4.4.

Biologically inherited developmental resources are a subset of parental resources. Cultural inheritance would include the inheritance of some parental resources and some collectively generated resources.

4.4 What Is the Challenge for a Concept of Human Nature?

An explanatory use of the concept of a human nature faces the developmentalist challenge because of the entanglement of nature and nurture.

This means, first, that if a trait is taken to be part of human nature because it is due to human nature, then such a claim needs to be consistent with the five dimensions of the interactionist consensus: genetic inertness, incommensurability, quantifiable difference making only, epigenetic interaction at the intergenerational level, and coevolutionary interaction at the evolutionary level.

Unfortunately, the prospects for there being a human nature, given the interactionist consensus, are dim: if "due to nature" means that nothing but nature in the sense of biologically inherited resources causally contributes to the trait, then there is simply no human nature in that sense because of genetic inertness.

Even if one restricts oneself to refer to difference making only (rather than causal contribution), the situation is desolate: to count a trait (such as language) to be part of the descriptive human nature since it is due to the explanatory human nature would require a genetic determinism. It would require that the trait has, for all possible worlds, a flat norm of reaction. In principle, this is possible. But it is very unlikely to hold for any important trait (e.g., rationality or language), given that even a simple trait such as body height does not have such a flat norm of reaction. The category of "traits that are part of human nature" is thus in danger of shrinking so much so that it is practically empty (i.e., so thin that there is nothing left). Given the interpretation of "something being due to nature'" as a "flat norm of reaction," human nature could exist, but likely only as a very thin oscillating shadow on the intellectual horizon with almost nothing in it.

Third, one can also challenge the divide between nature and culture by challenging the divide between biological inheritance and cultural inheritance, as suggested by developmental systems theory and interpretations of epigenetic inheritance. This is a challenge that mainly refers to interaction at the intergenerational and evolutionary levels. With respect to that, a defense of the concept of human nature that wants to use the nature-nurture divide has to give reasons for why it is justified to draw a line between nature and nurture (or culture more specifically) where it is traditionally drawn, namely, between biological inheritance and cultural inheritance. In addition, the defense should not metaphysically presuppose that the boundaries of the body allow drawing that line, for example, via intrinsicality as such, as in traditional essentialism. Developmental resources "traveling within" are traditionally and still today counted as biologically inherited and those "traveling without" (as in Galton's famous quote from above) are not. What is philosophically decisive is not that the line is drawn at the body boundary but why. Given that those resources traveling within and those traveling without can both be vertical (i.e., from parents to offspring) and effective over a long time, the justification cannot be that one of it (cultural inheritance) is only developmental. Biologically and culturally inherited developmental resources are both developmental and evolutionary. To stress this is a major achievement of developmental systems theory. In that sense, there is parity between these resources, as suggested by developmental systems theory. Yet by establishing this parity, one should not throw out the baby with the interactionist bathwater by suggesting to completely get rid of the nature-nurture divide.

In part II, I will stay neutral with respect to how thin or thick human nature is since that is an empirical issue. I will, rather, defend a version of the nature-nurture divide that refers to dynamic differences between biological inheritance and cultural inheritance. This defense, which opens a conceptual possibility, will take the worries I discussed as part of the developmentalist challenge into account. It will not rely on intrinsicality but on the third aspect of the traditional concept of human nature, fixity, which I will replace with stability. There is something that survives the developmentalist challenge.

Summary of Part I

The concept of human nature is under severe attack from a moral-political perspective and a scientific one. Part I systematized and discussed three challenges that the concept of human nature currently faces. The vernacular concept of human nature is perspectival and leads to ethically unjustified dehumanization of people who are considered as not being "fully human" (the dehumanization challenge). The more objective and scientific concept relies, as traditionally conceived, on an outdated essentialism that evolutionary thinking shows to be wrong (the Darwinian challenge). The concept in addition often relies on wrong assumptions about nature and nurture as explanatory categories: it is wrong to assume that genes (or all biologically inherited developmental resources) can do something on their own, that nature and nurture can be causally apportioned, that genetic determinism is widespread, and that culture acts only at the level of development (developmentalist challenge).

The overall conclusion is that the vernacular concept of human nature has a dark side, dehumanization, which is something that should be prevented. Human nature in that sense is vicious. The concept of human nature as traditionally conceived in sciences, providing an essence of what it means to be human, often connected with wrong beliefs about the causal importance of nature in development and evolution, is scientifically as outdated as the essentialism and nature-nurture divide assumed thereby. Human nature in that sense is dead.

While part I was critical, part II will be constructive: there are a couple of post-essentialist concepts of human nature that are here to stay despite the three challenges. Part III will then add some further normative issues and address whether one should try to eliminate any human nature talk, that is, whether one should try to eliminate the expression *human nature* to minimize dehumanization and prevent any association with outdated essentialist or dualist ideas.

II Three Natures: A Post-Essentialist, Pluralist, and Interactive Reply to the Three Challenges

As an exercise in population thinking, part II develops an answer to the three challenges from part I: it provides a post-essentialist, pluralist, and interactive reconstruction of human natures in the plural—a diversity of concepts of human nature that can be used in sciences and humanities despite the three challenges. It is a reconstruction of what "human nature" could still mean if one tries to keep dehumanization at bay and respects the constraints set by contemporary biological knowledge with respect to essences and the nature-nurture divide.

The starting point is to curtail the damage. Dehumanization reminds us of the danger in using the concept of human nature in a normative sense and the importance of objectivity. The Darwinian anti-essentialists debunked a specific definitional concept (the idea that life form features are necessary and sufficient conditions for species membership) and a specific explanatory concept (explaining variation away by reference to a normalizing essence hidden behind the variation). The interactionist consensus debunked genes as solitary producers, genetic determinism, and apportioning causality. Even the idea of separate channels of inheritance got questioned.

The guiding question for this part of the book is thus: Are objective, nonnormative alternatives for a concept of human nature available—alternatives that do not fall back to a dehumanizing, essentializing, and noninteractive (monocausal, deterministic, and too dualistic) use and still fulfill the epistemic roles that essences and the nature-nurture distinction traditionally fulfilled?

My answer, which I develop in this part, is that there are such alternatives. There is a classificatory nature, fulfilling the classificatory role; there is a descriptive nature, fulfilling a descriptive role; and there is an explanatory nature, serving a particular nonessentialist explanatory role.

None of these natures directly serves a normalizing and dehumanizing function and none of them is an essence in the traditional sense. This makes the account post-essentialist. The account is nonetheless realist since these natures exist objectively in the world. The account is pluralist since the scientifically important roles of traditional essences—the classificatory, the descriptive, and the explanatory—are each fulfilled by different things existing mind-independently in the world, things that can legitimately be called a "nature." The account is interactive since the interactionist consensus is taken seriously.

In sum, the message is that human nature in the essentialist sense is dead; long live the human natures in the post-essentialist and interactive plural.

The account treats genealogy as important for all three natures. Even though it holds that the three natures are distinct, they are connected by genealogy and therefore not completely independent. I take genealogy to refer to biological reproduction and therefore to biological ancestor-descendant lineages and inheritance. Genealogy is not only key for the classificatory nature (i.e., to determine the boundaries of species and membership of individual organisms in it); it also gives rise to the developmental resources that amount to an explanatory nature, which is of special relevance for the descriptive nature.

The most complex and controversial aspect of this revisionary picture is the reformulation of the nature-nurture dichotomy, at an evolutionary as well as developmental level. Discussing it in a critical but ultimately constructive manner will not only reformulate a nature–nurture (and nature–culture) distinction in a way that respects the interactionist consensus but also will lead to what I call explanatory looping effects. By giving explanations, humans are often already involved in creating the very human nature at issue in the explanations. In part, humans make human nature via their culture, not just symbolically but in reality.

I start in chapter 5 with the importance of genealogy: how it accounts for the classificatory nature and why it is indirectly, nonspecifically, and partially important for an explanatory and descriptive nature. Chapters 6 and 7 spell out what the descriptive nature is. The defense of nature as distinct from nurture or culture is part of chapters 5, 6, and 7. The different options for the respective explanatory nature are discussed in chapter 8. Finally, chapter 9 illustrates how human nature is in part created by explanatory looping effects.

5 Genealogy, the Classificatory Nature, and Channels of Inheritance

This chapter reconstructs the importance of genealogy for understanding the classificatory, explanatory, and descriptive nature of the species *H. sapiens*. Genealogy, as I understand it here, refers to biological reproduction and therefore to the intergenerational transmission of developmental resources and to the resulting biological ancestor-descendant lineages. The first main claim is that the genealogical nexus between humans is the classificatory nature, as Hull (1986) suggested. The second main claim is that genealogy gives rise to a specific channel of inheritance, biological inheritance, which bundles all those developmental resources that can be called an explanatory nature of humankind. The third main claim is that genealogy is thus partially and indirectly also explanatorily relevant for the descriptive nature, the life form of humankind, which amounts to a set of reliable descriptive generalizations about this group. This is how genealogy connects the three distinct post-essentialist natures of a species.

In section 5.1, I delineate five questions relevant for the discussion about a biological species' nature and then situate my account within a tradition following Hull, who defended genealogy as important for reflections about the nature of a biological species. I then illustrate how genealogy accounts for a post-essentialist classificatory nature by discussing in section 5.2 what genealogy offers for answering the first three questions: the species question, the constitution question, and the partaking question. Section 5.3 shows how genealogy gives rise to a specific channel of inheritance and how this provides options for a post-essentialist explanatory nature. In the final section 5.4, I outline the resulting pluralism. This includes a relativization of the importance of genealogy: whether genealogy is important in the specified manner results not only from how contemporary science works but also from terminological choices that depend on social experiences and values.

5.1 Five Questions Regarding a Species' Nature

When biological species are discussed from a philosophy of science point of view, five questions should be distinguished:

1. *Species question*: How can one individuate, that is, delimit a species? This is asking for a concept of species or taxonomic method. Instantiations of the question are, "What does it mean for a group of organisms to make up a species?" and, "How can species S1 be delimited from species S2?" (e.g., "How can *H. sapiens* be delimited from another species of the same genus?").
2. *Constitution question*: How are species evolutionarily constituted, that is, how is the homeostasis of the historically changing property cluster characteristic for a species (and the corresponding generalizations) made possible? This is asking for what I call a constitution explanation. Instantiations are, "How do traits of a species persist and cluster over evolutionary time?" and "What are the general mechanisms of species cohesion?"
3. *Partaking question*: How can it be decided who belongs to the species? This is asking for a sorting decision. In other words, it relates to the extension of the kind term. Instantiations are, "Who or what is a member of S1?" and "Why is organism O a member of species S1 rather than S2?" (e.g., "Why does Tiggy belong to tigerkind?").
4. *Description question:* How can a species be described qualitatively? This is asking for a description of the property cluster, that is, the life form of a species: typical and stable (maybe even specific) characteristics of the group that counts as a species. Linguistically, it targets the intension of the kind term. Instantiations of the related questions are, "What is it like to be a member of S1?" (e.g., What is it like to be a human?), "What does it mean for O to be a member of species S1?" (e.g., What does it mean to be human?)
5. *Trait explanation question*: How are the evolution and reliable development of particular traits of the life form of a species in an individual or the group explained? This is asking for an evolutionary or developmental explanation of specific traits in a particular species. Instantiations

are, "Why do members of the species S1 have property P?" (e.g., "Why do tigers have stripes?").[1]

The species, partaking, and description questions are not asking for an explanation. Only the constitution and the trait explanation questions ask for an explanation. Furthermore, the species, constitution, and partaking questions refer to species in general. The answers consequently hold for species in general. By contrast, the description and trait explanation questions refer to particular species. Certainly, the first three questions can have concrete applications to particular species (e.g., "Where are the boundaries of species S1?" "What constitutes the homeostasis of species S1?" "What are the membership conditions of S1?"), but they are primarily general questions about species as a category. Consequently, answers to them will apply to species in general. We thus have three so-called species category questions (concerning species as such, as category, questions 1 to 3) and two species taxon questions (concerning particular species, questions 4 and 5).

Distinguishing among the five questions is important since they are too easily conflated. For instance, if talk about the nature of a species is taken to be about "what it is to be" a member of species S, recurring frequently in Devitt (2008), then this is too unspecific.[2] If this question is further specified as, "In virtue of what is an organism an F? What makes an organism an F? What is the nature of being F? What is the essence of being an F?" (Devitt 2008, 357), then nothing has been gained. Although these formulations are all distinct from the species question (which Devitt calls the "species category question"), they do not allow distinguishing among the other four questions. An undifferentiated way of discussing what it means to talk of the nature of a species is often triggering the monism that stands behind traditional essentialism: it prevents one from seeing that the answers to the different questions refer to different things in the world.

Three Claims Based on Distinguishing among the Five Questions
On the basis of distinguishing these five questions, I defend three claims. First, there are three separate natures: a classificatory nature that allows replying to the species and partaking question, namely, the genealogical nexus; a descriptive nature, namely, the stable phenotypic life form of a species, which allows a reply to the description question; and an explanatory nature, namely, the species' pool of biologically inherited developmental

resources, which allows us to give a partial reply to the constitution and the trait explanation question. Only the classificatory nature can be fully described in this chapter.

Second, genealogy provides the foundation for distinguishing between biological inheritance and cultural inheritance, and thus between nature and culture. To be able to distinguish between the two is a necessary requirement to specify the explanatory nature as suggested, that is, as the pool of biologically inherited developmental resources.

Third, genealogy connects the classificatory, the explanatory, and the descriptive nature by being relevant for all of them. Genealogy is the frame for all the three natures, which are nonetheless not the same. That genealogy provides a connection also means that it is relevant for all five questions. As long as partial and indirect explanations can be explanations, genealogy is classificatorily relevant for the species and the partaking questions and explanatorily relevant for the constitution, the description, and the trait explanation questions.

About Species, Not Biological Kinds or Organisms

It is important to note that the five questions are about biological species, not biological kinds in general. The terms *organism* and *predator* refer to biological kinds but not to species. Consequently, the importance of genealogy to answer these questions holds only for the species category. Sober (1984) and Wilson, Barker, and Brigandt (2007) claim that there are biological kinds, such as the kind "predator," that do not form a lineage of descent. Consequently, genealogy will be of less or even no importance to answer respective questions about the nature of such a kind.

This is important, since (as indicated earlier) the term *species* is quite contested. I use it as contrasted with *kinds of organisms*. I thus use it for a special sort of a biological kind, namely, for an *evolutionary kind*. In the background is the following. Dumsday (2012) suggested that there are two ways of classifying biological individuals, depending on disciplinary interest: as groups of organisms and as species. If classified as groups of organisms (e.g., in the laboratory), he argues, entities necessarily have, in part, an intrinsic essence. I agree, but I add that if the entities are classified as species (as in phylogenetics or evolutionary theory), then the same individual entities equally necessarily have in part a relational essence since species as evolutionary units involve populational lineages, that is, genealogical relations between organisms. With this I am not excluding nonevolutionary uses of the term *species*, as in specific contexts such as biodiversity research. I am only saying that in evolutionary contexts, lineages are definitional for species. I further defend this in section 5.2.

A Tradition from Hull's Genealogical Account
A tradition descending from Hull already takes genealogy very seriously for answering questions about the nature of a species, even though each author does so in slightly different ways. I call them taxonomic relationalists. Millikan (1999), Griffiths (1999), Okasha, (2002), and Rieppel (2009, 2013) belong to that group since they defend a relational account of the essences of species. They take the life form of a species and also developmental resources and mechanisms that are explanatorily relevant for the life form to be anchored in genealogical relationships (in the case of Okasha, in other relations as well). The classificatory essence of a species consists in relations between individuals partaking in the species rather than in the similarity of intrinsic properties of these individuals. In their account, similarity is regarded as derivative only. They all grant that using a field guide is often all one can do to classify individuals, since genealogy is often unknown or even unknowable from the standpoint of those who have to classify individuals. Yet the field guide (and thus the similarities in it) are used as a proxy only, as an epistemic indicator, not as a logical criterion for the species' boundaries and membership in it.

As long as relations are primary and trump similarity, these approaches are compatible with Boyd's (1991, 1999) account that takes homeostatic property clusters as the post-essentialist successor notion for traditional essences, providing simultaneously a definitional as well as an explanatory and descriptive essence. Properties hang together, according to such an account, because of so-called homeostatic mechanisms, which are "mechanisms of causal integration that endow species with a 'nature,'" as Rieppel (2013, 165) summarizes the account. The compatibility derives from the openness of the concept of homeostatic mechanisms since these can include the relations that the taxonomic relationalists consider as key for the species question and the partaking question. Thus, taxonomic relationalists can be taken to defend a specific version of a broadly conceived homeostatic property cluster account.

I develop an account in the spirit of these relational accounts by distinguishing more decisively than so far among the five different questions, which allows a more pluralist post-essentialism: classificatory, descriptive, as well as explanatory nature as existing but falling apart. Except for Okasha, the taxonomic relationalists seem to assume (as traditional essentialists do) that there should be only *one thing* replacing traditional essences. While some try to recover more than one traditional epistemic role of the concept of an essence, Okasha is monist in a different sense: he claims that there is only *one epistemic role*, the classificatory role of the one concept of a species' essence, that can and needs to be rescued from the Darwinian challenge. It is all one can have, but as he seems to assume, luckily also the only role a taxonomic scientist needs. That might well hold for the taxonomist, but there are other scientists using the concept of a species' nature. Although I agree with Okasha on what he says about the classificatory role, the needs of other scientific fields need to be taken into account too. This will also show that genealogy is ultimately important across scientific fields and even in society.

5.2   Genealogical Nexus as the Classificatory Nature

The Species Question

Genealogy answers the species question, though only in part (i.e., together with a speciation event). A species (in the unified definition of Queiroz 2007, 880) is a "separately evolving metapopulation lineage." Species require breaks in continuity of lineages and are thus defined by the distance in genealogy that mirrors the separateness of their evolution. The temporal boundaries of such genealogically defined species are determined by speciation events, which are instantaneous only from the perspective of the life span of a species. For our (after all diachronically very myopic) science, a speciation event will probably be forever unobservable and vague in terms of which individual or group was the first of its kind. Thus, where exactly the boundaries of a species lie will forever be underdetermined by the sheer vagueness of the matter. Furthermore, whether a speciation event happened is dependent on what happened after it. It can thus only be established vaguely and after the fact.

Nonetheless, scientists can use evidence to model or infer what happened—evidence from interbreeding, geographic barriers, and niche construction. Which evidential criteria allow (or allow best) the conclusion that a speciation event happened, roughly here or there, is thus a separate issue. Taking genealogy as a logical criterion for what a speciation event is—the emergence of genealogical separateness—is independent of how we know about it.

Nonetheless, the evidential criteria are of utmost importance for the methods in biological taxonomy and for the concrete conclusions drawn, for instance, in paleoanthropology. A lot of disagreement exists regarding them.[3] That there is no consensus regarding how to rank these criteria (especially in cases where the evidence conflicts by pointing in different directions) is creating further underdetermination of any claim about the origin or end of a species. For instance, the evidence for when the species *H. sapiens* and the genus *Homo* begins seems to vary from discipline to discipline (e.g., from paleoanthropology to molecular anthropology). To the best of my knowledge, no consensus about the right conclusions exists so far.

In addition, social values, especially in the case of humans, might weigh the evidence in biased ways. Proctor (2003) has suggested one such influence in his work on the history of claims about human origins. He states that whether claims about the origin (i.e., "birth") of humans tend toward more antiquity of humans or more recency also depends on moral and political choices, for instance, on how antiracist the society is in which the scientists who discuss claims of that sort are situated. Depending on whether otherness is regarded as something to be highlighted (rather than downplayed), more groups are excluded (rather than included), synchronically in society as well as diachronically in the species. How much interbreeding is enough to count populations as still one species? This is just one of the questions that involve an underdetermination that stems from the vagueness in "how much" thresholds that open the gate for values to influence the decision about how much is enough. The more exclusive "we" are, the younger "we" become; the more inclusive "we" are, the older "we" become. For instance, if it is concluded (influenced by societal changes) that new scientific evidence about some interbreeding is enough to conclude that Neanderthals were not a separate species (and also that the differences between these *H. sapiens-neanderthalensis* and the *H. heidelbergensis* are not enough to justify regarding them as separate species), then "we" (those so far regarded as non-Neanderthals and non-Heidelbergensis) get much older, together with the others. There is then only one united and older species of *H. sapiens* (rather than separate *H. sapiens*, *H. neanderthalensis*, and *H. heidelbergensis*). Its life started when that inclusive species split off from another still older species, most often taken to be *H. erectus*. Contemporary overviews about dating human groups therefore include error bars, reflecting "various sources of errors," as Wood and Boyle (2017, 24) write. This fits what Proctor claims: in the post–World War II climate (mainly because of the Nazi ideology and related atrocities), there was a tendency to push our origin backward that stemmed from the social tendency to be more inclusive (after having been very exclusive), a tendency that seems to be fading away once again. How old the human species is judged to be is also depending on what Proctor calls the "discovery/reward process" of science. After all, a scientist will not get fame or a lot of research funds if, while searching the fossil fields at the boundary lines of our species or our genus, she declares that the precious fossil she found is the last nonhuman primate. But she will get a lot of fame (and eventually funds) if the fossil can be labeled as the first human. Independent of whether Proctor's specific claims are correct, his work illustrates how values can influence claims about speciation.

It is complicating matters that some even consider those criteria that Queiroz regards as merely evidential, equally definitional, or solely definitional. Which niche a species has is, for instance, taken to determine species boundaries as part of the so-called ecological species concept. If these evidential indicators are taken as definitional, then genealogy can even become secondary and indicative only. Species would then be defined in a completely synchronic manner, without any historical dimension. Ecology might well be in need of such a species concept. Yet this already indicates that there might be different disciplinary needs involved. Ecologists treat organisms as species but not as evolutionary kinds, that is, not with the same historical dimension as phylogeneticists. It is the plurality of contexts (each with its own goals for explanation: explaining evolution, explaining ecological adaptation, explaining functioning, and so on) that has led to a pluralism regarding species concepts, defended, for instance, by Kitcher (1984), Dupré (1993), Queiroz (2007), and R. A. Richards (2010).

Although I agree with this pluralist camp, I defend that there is a shared thread in the plurality, a thread that is explored in detail by Richards's (2010) pluralist account. Whatever context is chosen, the following still holds: as evolutionary units, units of a diachronic process, species are genealogically separate lineages of metapopulations. If a species concept does not involve a genealogical dimension, the resulting species concept is not a useful taxonomic unit for evolutionary theory—even if it might be a wonderful ecological unit or unit of adaptation.[4]

As long as species are understood as evolutionary units, genealogy is necessarily part of the answer to the species question. That genealogy is not sufficient and that there is no consensus which additional criteria are—together with genealogy—sufficient is an important issue but irrelevant for this study.

Constitution Question
The evidential criteria that one can use to delimit the boundaries of a concrete species are also explanatory but with respect to the constitution question. They explain why populations are moving apart or not. Even if what it means to be distinct is first and foremost determined by genealogy, not by the evidential criteria, the latter refer to homeostatic mechanisms that keep organisms together or apart. Interbreeding, geographic barriers, and niche construction in the earlier phases of speciation, and genetic and developmental constraints in the later phases, not only keep groups apart; they also keep members of one species together by keeping developmental resources and thus the resulting traits together. These explanatory factors (some of which are processes, some structures; call them mechanisms, if you like) cause the homeostasis, the stability of the properties that are characteristic for the respective species, a stability in place (synchronic coherence) and time (diachronic stasis).

This set of factors explains the homeostasis of species in general, although for each particular species, a different subset of them (and concrete instantiations of the relevant factors) will explain both the synchronic coherence of the properties of that particular group and the relative diachronic stasis of that particular group with its property cluster.[5]

To sum up, homeostatic mechanisms explain how and why a property cluster clusters (synchronically as well as diachronically), that is, how and why the observable properties used in the description of a species glue together. They thus provide an answer to the constitution question in its general form, whereas each particularized constitution question will refer to specific homeostatic mechanisms.

Partaking Question and the Resulting Classificatory Nature
Genealogy answers the partaking question in all practically relevant situations once there is a clearly delineated species, since only then does the question make sense at all.

The classificatory human nature that results can be specified in the following way:

> Classificatory nature = the genealogical nexus, i.e. (if one wants to call it a property) the relational property of being genealogically related to (being a descendant of) other humans. This nexus is necessary and sufficient to belong to the species *H. sapiens*.

The genealogical nexus exists mind-independently and whether two organisms are genealogically related can be objectively established. I thus take it as an established scientific fact that all conventionally recognized contemporary human populations are genealogically connected and in that sense one kind, one species, one "family," as the UN Declaration of Human Rights states. I also assume that this genealogical concept of human nature—and only derivatively any descriptive knowledge about the human life form—contemporarily lies at the foundation of the universalism in the idea of one humanity, which got instantiated in the various twentieth-century declarations of human rights.[6]

If one were using essence talk (which I recommend not to), the genealogical property of "being genealogically related to humans" would be the definitional essence of humankind. But it is (if at all) a relational property, not an essence; it only performs one of the epistemic roles of a traditional essence, namely classification. For this reason I call it a classificatory nature.

The neo-essentialist Devitt (2008) also admits that an organism O is a member of the species F if and only if it is in a genealogical relationship to other Fs. Yet he claims that this does not "tell us what property makes an organism a member of the group of *F's in particular*." He does so in order to argue against the tradition descending from Hull and by using a point that is often made: that the genealogical relation is transitive. If A is genealogically related to B and B is genealogically related to C, then A is genealogically related to C, and ad infinitum. Ultimately a member of *H. sapiens* is thus genealogically related to members of different species, and also to those entities at the beginning of life. Yet this transitivity argument (as I call it) ignores that Hull (1986) already replied, *avant la lettre*, more or less in the following manner: if an individual is genealogically related to Fs, it is directly related to Fs in particular, and only indirectly and in quite nonparticular ways to members of other species on the tree of life. It is the direct genealogical relation that is used to answer the partaking question. Hull's solution thus survives.

It survives even though the partaking question makes sense only in situations where there are Fs, that is, a species discernable regarding which the partaking question can be asked in meaningful ways. An ethically relevant situation of asking the question is, for instance, one where one needs to decide whether one is morally obliged to help (an individual that is in need) in specific human ways (e.g., a refugee drowning in the course of fleeing from war or other hardships). In such a practically relevant situation, there is already a taxonomically clearly delineated kind (*H. sapiens*), and the partaking question can be decided by genealogy. Is that individual *H. sapiens*: yes or no? Genealogy as the logical criterion will decide the question, even if nobody will ask for a birth certificate or any other test for membership in humankind. The look (an evidential criterion) will be sufficient. In addition, the respective individual is as directly *H. sapiens* (or not) as I am directly the child of my specific parents (or not), even though I am certainly also genealogically related to more distant ancestors. The logic of transitive relations won't ever change that there are direct and indirect genealogical relations.

The major problem is not that there is a transitive relation involved (since it can be solved by distinguishing the partaking question from the other questions and by using the pragmatic argument just outlined). The problem is rather that the procedure has a smack of tautology, as already mentioned in chapter 3. To answer the question, "Who is human?" by saying "Those who descend from humans!" is semantically close to a tautology. Nevertheless, it does, pragmatically viewed, its job: answering the partaking question in all relevant situations, that is, in all situations in which there is a species. In other situations, when there is no clearly delineated species yet, nothing except recourse to the species question might help, which is what those asking for when "they" became "us" are doing.[7]

It should be clear that genealogy as a criterion for species membership is rather nonoperational. I alluded to this when I discussed that the look of refugees is enough to grant them the right to be saved from drowning.

But as Okasha (2002, 195, 202–204) wrote, with reference to Kripke (1972) and Putnam's (1975) traditional accounts of essences as hidden, there is no good argument for why operationalism should be assumed, that is, why the nature of a kind needs to be observable. Speakers in science and outside are able to have an implicit agreement that their kind terms rely on a hidden, mostly unobservable classificatory nature. In addition, they can use observable indicators (i.e., epistemic identifiers) for the hidden genealogical nexus. The list of epistemic identifiers is something like a field guide. Thus, one does not need to ask for birth certificates to classify somebody as *H. sapiens*. But I assume that if one would find out that an organism that one regards as *H. sapiens* is not conceived by *H. sapiens*, then most would stop regarding the individual as a member of the species *H. sapiens*. We would call it a humanoid, maybe even a human, but not a member of *H. sapiens*. Furthermore, in most contexts, biologists equally do not literally see genealogy but morphology (i.e., similarities in the traits the organisms exhibit). They then classify individuals according to the observed similarities. The important question is, however, whether they treat similarity as proxy, as reliable indicator, as evidence for genealogy, or as an independent criterion. I assume that most biologists today would agree that in evolutionary-taxonomic contexts, similarity is only the proxy for genealogy. Genealogy is what one aims at, even if one cannot see it.[8]

5.3 Genealogy and the Channels of Inheritance

Before I can move on to the description and trait explanation question and thus to the other two natures that survive the three challenges from part I, the connection between genealogy and the nature–nurture (and nature–culture) divide needs to be specified.

Genealogy enters the answer to the constitution question since it circumscribes a peculiar channel of transmission of developmental resources. Those phenotypic properties that are held together, as well as the mechanisms that keep them together, have to endure in order to cause stasis (i.e., diachronic coherence) as well as speciation. Yet resources that influence development and evolution can endure in different ways: by persistence, biological inheritance, or social learning. Geographic barriers such as mountains separate populations and are thus part of the homeostatic mechanisms causing stasis as well as speciation, but they are not inherited;

they simply persist. Resources important for stasis and speciation that do not simply persist (once they are there) have to be transmitted by duplication or re-creation in the next generation. Transmission can happen by genealogy (i.e., biological reproduction), as it does for genes and other parts of cells, or by social learning, as it does for habits, rights, or knowledge.

In other words, for the receiving organism, there is nature (transmission via biological reproduction), there is culture (transmission via social learning), and there is persistence of ecological factors. This makes three main channels of inheritance of developmental resources.

Chapter 4 showed that this channelism is contested because of the interactions among the different channels. The central claim of this section is that the biological channel can be dynamically delineated from culture, the other transmission channel: whether transmission happens via nature or culture makes a difference for the overall homeostatic dynamic, and it is this dynamic difference that allows delineating the two transmission channels. That one can distinguish between transmission and persistence will not be analyzed any further; it will be taken for granted in the following.

I proceed by introducing the distinction between biological reproduction and social learning as an age-old, commonsense distinction of different kinds of social interactions, each connected with a distinct intergenerational process forming a channel of inheritance. I then present three criteria that help with "digging the channels"—help with demarcating separate sets of developmental resources traveling from organism to organism via distinct channels of inheritance. These three criteria are the autonomy of cultural inheritance mentioned in chapter 4, near-decomposability, and a difference in temporal order. These three points not only make the distinction between biological and cultural inheritance more precise; they also justify it as real, that is, ontologically adequate. Yet most important for the aims in this chapter is the difference in temporal order, since it shows that transmission in the cultural channel is content dependent, an important cause for why culture contributes in a much less stable manner to the homeostasis of a species.

Biological and Cultural Inheritance as Distinct

As mentioned, there are different ways in which developmental resources endure over time. Environmental resources simply persist; they are not transmitted or replicated from organism to organism. With respect to transmitted resources, it is common sense to distinguish between biologically inherited ones and culturally inherited ones. Biological inheritance happens via biological reproduction, whereas cultural inheritance happens via social learning.

That biological reproduction and social learning are distinct is common sense since, after all, sex (as one way of biological reproduction) is simply not the same as talking (to take one kind of social interaction that can lead to social learning), even though sex usually involves some social interaction, if not social learning. In a nutshell, biological reproduction and social learning simply refer to different types of interactions between people, with different characteristics. For instance, biological reproduction regularly involves material overlap (especially for the females) and normally occurs within bodies, since it ultimately relies on (meiotic) cell divisions.[9] The transmission of developmental resources via social learning normally occurs outside the body and only sometimes involves handover of material entities. If it involves material entities, this usually makes the transmission more reliable, as when scientific results are published in print or when a contract for legal inheritance of property rights is printed in order to make the transmission of the respective content or property more reliable.

In the following, I show how the two kinds of interactions, biological reproduction and social learning, the two channels of inheritance in focus in this chapter, can be more formally delineated.

The autonomy and near-decomposabililty of cultural inheritance Given that Lamarckian inheritance is not possible, cultural inheritance can take off in the sense displayed in chapter 4. Cultural inheritance is autonomous, meaning that it is decoupled from biological inheritance. There are then two separate and decoupled types of interactions between organisms, building two separate channels of inheritance regularly reoccurring over evolutionary time in empirically discernable ways. The boundaries of these two different channels of inheritance can be established by the principle of near-decomposability. Merlin (2010) observes that given specific channels of inheritance, the "interactions among their respective own elements" are "stronger than the interactions among elements belonging to different subsystems" (i.e., channels) of the overall inherited developmental system.

Since Simon (1962), such cases of more interaction within than between subsystems are called cases of near-decomposability. If there is near-decomposability in a whole, then it is safe to draw an ontological boundary between the parts of the bigger whole.

I cannot think of anybody who seriously wants to deny that genetic and other cell-level developmental resources interact more with each other than with developmental resources that are clearly culturally transmitted (e.g., spoken words). Take the case of the evolution of lactose tolerance, a paradigmatic example of gene–culture coevolution. The dairy farmers in the Levant and then Europe who changed the prospects of the selection of the genetic change that made lactose tolerance in adults possible, had daily interactions to produce milk and then, for instance, to use and produce the pottery that allowed them to make cheese, pottery that is now archaeologically studied to track the coevolution of dairy farming and lactose tolerance. None of the genetic factors making lactose tolerance possible was causally directly involved in that daily business of dairy farming, since the genes involved are not genes for dairy farming. Thus, it is safe to assume that the principle of near-decomposability applies to the case at issue and to cases of coevolution generally.[10]

Near-decomposability as a dynamic way of drawing system boundaries applies to all kinds of entities and is in that sense formal and ontologically maximally neutral. For instance, it is usually also safe to consider departments as parts of universities, since there are elements (members of the universities such as professors, staff, and students) that interact more within individual departments than between different departments. In the same way as there are then such departments of universities, there are channels of inheritance: inheritance (analogous to a university) has parts, namely, different subsystems (analogous to the departments), whose developmental resources (analogous to the members of the university) interact more with each other than with the developmental resources from other subsystems. A channel does not exist as a material entity; it is not a barrier preventing equally distributed interaction of the elements, like a wall or tube, although it can certainly involve such material barriers (e.g., as bodies separate people and buildings departments). A channel is not a material thing, but it depicts a real pattern—one of causal interactions and differential distribution of causal interactions that allows drawing a more-within-than-between boundary of parts of a bigger whole. Calling the channel a "frame" is probably the best metaphor: it highlights which part of a "picture" is a proper part.

As long as it can safely be assumed that near-decomposability applies to the distinction between biological reproduction and social learning, the biological channel can be delineated from the cultural channel. The commonsense distinction between biological reproduction and social learning, together with the autonomy and near-decomposability of the channels, suffices to establish the distinction between biological and cultural inheritance as real, that is, as cutting reality at one of its joints. Yet there is a third reason for it: a further dynamic difference, namely, a key difference in temporal order over evolutionary time, a difference that connects back to the autonomy of cultural inheritance from biological inheritance.

Content-independent and content-dependent modes of transmission The mode of transmission in the biological channel is mostly content independent, whereas the transmission in the cultural channel is content dependent. Depending on content, the mode can change, say, from vertical to horizontal. After describing how modes of transmission can be vertical, horizontal, or oblique, I will defend the claim that there is a difference with respect to content dependence of transmission and show that this has consequences for the stability of the transmitted developmental resources.

Cultural inheritance can be divided into modes of cultural transmission, such as vertical versus horizontal transmission, as originally proposed by Cavalli-Sforza and Feldman (1981, 54). Vertical cultural transmission runs parallel to genealogy, from parents to offspring. Horizontal cultural transmission is between any unrelated individuals. If that transmission is between individuals from different generations, then it is oblique. Cavalli-Sforza (2000, 179–187) later refined this tripartite taxonomy of cultural modes of inheritance into a more fine-grained grid of four modes: a vertical mode plus three different kinds of horizontal modes of transmission, the latter distinguished according to the number of senders and receivers involved. There is a one-to-one communication mode, as in peer-to-peer interaction. A magistral transmission mode, by contrast, is a one-to-many communication pattern, as in mass media or cases where an authority enforces a cultural item on a population by decree, political or social pressures, or

Genealogy and Channels of Inheritance 107

simply network power. There is also a concerted transmission mode, which involves a many-to-one communication, where a group of people exerts social pressure on new members, as in religious cults. The four modes of cultural inheritance can be further divided into transmission processes that exhibit some bias (e.g., "copy-the-best," "copy-the-neighbor," in which learning is biased toward certain senders, e.g., one's neighbor). All of these are distinctions regarding modes of transmission within the subsystem of inheritance usually called culture.

The different transmission modes of culture all show a different temporal order in terms of kind of change (how a new item in the population spreads, from many to one, one to many, many to many, and so forth) and stability (how quickly things change). Vertical transmission is usually very conservative, though less so than concerted transmission; magistral transmission is very handy for quick and reliable spread of novelty, which is why it is used in schools and by governments. Yet it is less creative and slower compared to peer-to-peer learning, as fast-changing dress codes or any game of telephone makes clear.

The actual transmission dynamics in concrete cases are quite complex and depend on a lot of contextual variables. Claidière and André (2012) therefore criticize the assumption that the temporal order (how and how quickly a change happens) in cultural inheritance is a result of different modes of transmission. They see it as the other way around: the transmission modes are the result of transmission processes that are determined not by different modes but by the content of what is transmitted and the preferences of the human beings transmitting and thereby selecting cultural items. If so, then the alleged modes of transmission are real but only as epiphenomena in the sense that they do not exist as mechanisms for transmission in the world. They are observable, but they are merely a result of the diversity of the actual transmission processes.

A core argument for this claim is that a cultural trait (e.g., the concrete instantiation of the habit to cook food in a specific way) can change its mode of transmission from vertical to horizontal. Imagine the following. The availability of water changes, and you learned from your parents how to prepare potatoes by cooking them in water. Given the new evolutionary affordances, you learn from your friends, who, in contrast to your parents, quickly adapt to the new evolutionary affordances and bake the potatoes in the oven without using water. The actual content of the item transmitted (how to prepare potatoes) and the evolutionary affordances (whether adaptation is quickly needed or not) determine whether it makes sense or not what your friends and parents do and thus whether the mode in which the information travels is fast track (horizontal transmission) or standard security track (vertical transmission).

By contrast, what is biologically inherited normally cannot change their mode of transmission just because their content in relation to evolutionary affordances would suggest that.[11] Mitochondrial DNA is transmitted maternally, whatever it does in its context; nuclear DNA is (in sexually reproducing species) transmitted biparentally by meiosis; methylation patterns are transmitted epigenetically; and other cellular resources are transmitted at their cellular level. The mode of transmission in these cases of biological transmission is independent of the content of the item being transmitted and thus is also independent of evolutionary affordances, at least for a very long time (since certainly the way these molecules are transmitted ultimately evolves in relation to evolutionary affordances). The mode of transmission depends on the kind of molecule transmitted. It is mechanically fixed by the material constitution and the mechanisms of biochemical reactions, making inheritance possible for such a kind of molecule. Finally, most of the mechanisms involved in biological inheritance are vertical, with the exception of lateral gene transfer, which rarely occurs for multicellular organisms (on current evidence).[12]

What follows from this for the task of delineating biological and cultural inheritance? Claidière and André (2012) have used the argument to question the usefulness of distinguishing between modes of inheritance within the channel of cultural inheritance. Yet their argument (though only implicitly and, I guess, unintentionally) strengthened the divide between cultural inheritance and biological inheritance as operating in dynamically distinct ways. If they are right, there are two quite distinctive ways in which developmental resources can be transmitted over time: in a mostly content-independent manner (nature) and in a content-dependent manner (culture).

This difference has important consequences for the stability of inheritance. Over an evolutionary timescale, the two channels show a different temporal order with respect to the stability they convey for the reoccurrence of the respective developmental resources across generations. The biological channel leads on average to much more stability since the resources traveling within it normally cannot switch its mode of inheritance and thus stay predominantly vertical, a mode that has, on average, a high fidelity and thus stability. Culture, by contrast, is normally dependent on the choices of humans who react to the contents of culture and context-dependent affordances. These choices make culture much freer, less predictable and less secure in terms of stability.[13] This fits Danchin et al.'s (2011) distinction among genetic, epigenetic, parental, ecological, and cultural transmission. They characterize the temporal order of the different channels that they distinguish by pointing to vertical versus nonvertical transmission modes (see figure 5.1).

Even given nonvertical (i.e., lateral) gene transfer (the two small arrows in the field of genetic inheritance), the figure illustrates that cultural inheritance is quite distinct and special in terms of the temporal order it has since no mode dominates.[14] With Claidière and André (2012), one can explain why. In cultural inheritance, no mode dominates since transmission is

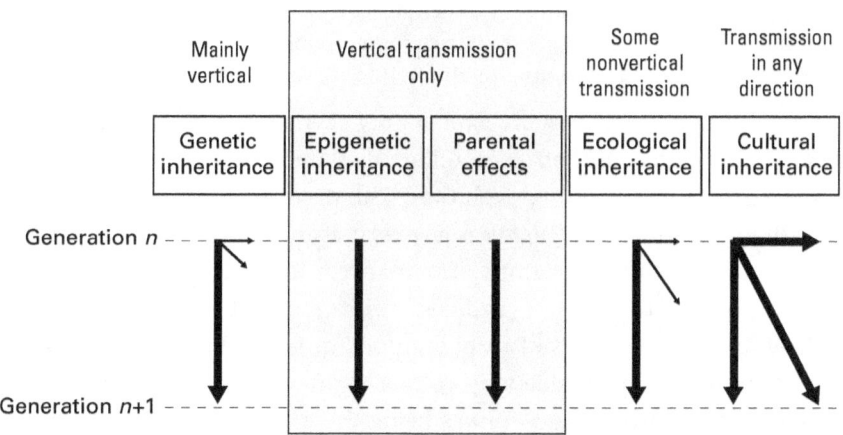

Figure 5.1
Channels of transmission. The channels of inheritance are characterized as vertical, horizontal, and oblique transmission modes, represented as arrows pointing vertically, obliquely to the right, and horizontally. (Reprinted by permission from Macmillan Publishers Ltd: Nature Review Genetics; Étienne Danchin, Anne Charmantier, Frances A. Champagne, Alex Mesoudi, Benoit Pujol, and Simon Blanchet, "Beyond DNA: Integrating Inclusive Inheritance into an Extended Theory of Evolution," *Nature Reviews Genetics* 12, no. 7 [2011]: 475–486, 481, © Macmillan Publishers Ltd 2011.)

content dependent. Mode of transmission thus changes all the time, not just from one cultural trait to another (e.g., from religion to dress codes) but also for a particular cultural content (e.g., how potatoes are prepared). The dynamic characteristics of culture depend on lots of contextual variables, since these decide whether transmission is vertical, horizontal, or oblique.

The fact that biologically inherited resources are bound to a mostly vertical mode of transmission is one of the reasons why the biological channel is regarded as characterized by, on average, higher stability. It therefore can also be regarded as exhibiting—again, on average—a greater evolutionary scope. I take the concept of evolutionary scope from Merlin (2010), who states that channels (or systems, or modes) of inheritance "empirically differ in important respects, in particular with regard to the scope of their effects across evolutionary time." She argues that

> the evolutionary potential of hereditary factors transmitted ... over many generations *is more important* than the evolutionary potential of hereditary factors transmitted over only one generation because the former, by being transmitted more or less reliably across many generations, can be subject to several events of selection and/or drift, and so can potentially lead up to more significant long-term evolutionary consequences (e.g., adaptations). (Merlin 2010, 212, emphasis in the original)

This, she says, establishes an "explanatory asymmetry" regarding the different subsystems of inheritance: different channels have different explanatory power for explaining evolution. The more stability inherent in an inheritance channel, the higher is its explanatory power.

The specific feature of cultural inheritance is that its scope is not predictable, in contrast to the quite predictable great scope of the biological channel. With this comparative lack of temporal order in the cultural channel, it also becomes explainable why culture is often treated as being able to change more quickly—and even as being there (having evolved), because it can change quickly if necessary. Certainly culture can also be quite conservative (e.g., if it is vertical). Yet culture is, as Alfred L. Kroeber (1917) already assumed a hundred years ago, unique in being nonetheless the perfect fast track of evolution. Jablonka and Lamb (2005, 298) therefore speak of culture being in "the driver's seat." The decisive point in this chapter is that it depends on the context whether culture is conservative or flexible, not on the characteristics of the channel. The channel allows for both flexibility and stability. The scope is thus not predictable from the channel itself. Knowing that a resource is transmitted from individual to individual via social learning rather than biological reproduction does not mean that one learns anything about its dynamics.

By contrast, the genetic mode—until technological solutions are available—does not allow for a similar fast-track mode and the resulting flexible changes everyone knows from culture. That is even the point of it: it is the function of the genetic mode in the evolution of multicellular organisms to have high stability.

Epigenetic Inheritance, Niche Inheritance, and a Diversity of Partitions
Before I can conclude this section that aims to show that genealogy is giving rise to biological inheritance, I need to add some further remarks on epigenetic inheritance, niche inheritance, and the diversity of partitions between channels in the current literature.

The epigenetic mode of biological inheritance (in the narrow sense as introduced in chapter 4) is a mode within the channel of biological reproduction. It is not part of cultural inheritance since the transmission of epigenetic variation happens at the level of cells that are inherited through biological reproduction. In addition, as described in chapter 4, parental effects in and of themselves are not cases of transmission but cases of social interaction involving a kind of embodiment that is not specific to epigenetics. It can lead to inheritance, but then it leads to either a standard epigenetic inheritance, which is biological and not cultural, or simply to cultural inheritance as traditionally conceived. Thus, parental effects cannot challenge channelism—the claim that there are sets of explanatory factors traveling separate channels of inheritance.

Epigenetic inheritance is there to fine-tune the biologically transmitted developmental resources. It operates within a few generations only, despite being intergenerational and vertical. Jablonka and Raz (2009) quote evidence that in the case of humans, the epigenetic marks last only up to three generations. There is thus some evolutionary scope of epigenetic inheritance, even for humans, but it is quite limited. Culture, by contrast, can have an enormous scope (even if it does not have to). In sum, the biological channel keeps a ledger of all those developmental resources that should persist for a very long time in complex systems such as organisms. To keep this ledger successfully, some fine-tuning via epigenetics seems to be adaptive.

The claim about higher stability of the biological channel thus holds even if epigenetic inheritance is taken into account: the overall biological channel exhibits on average (taking genetic as well as epigenetic inheritance into account) a higher stability than the cultural channel. Cultural inheritance, as a channel of transmission, is possibly a faster channel, but it is therefore also a less secure channel for creating stability over time.

Niche inheritance is another case that might be regarded as conflicting with the claims of this section, but it does not. Odling-Smee (2007, 2010) regards cultural inheritance as part of what he calls "niche inheritance," the inheritance of all developmental resources that travel outside the respective organisms and therefore only indirectly from ancestor to descendant. A niche is a set of environmental resources for development that either persist (e.g., the mountain) or need to be duplicated and recreated for each generation (e.g., a nest). In Odling-Smee's picture, the biological channel is also treated as distinct, but it is contrasted with the channel of niche inheritance, which includes cultural inheritance and ecological persistence, cultural inheritance being "the primary means by which humans engage in the universal process of niche construction," as Odling-Smee and Laland (2011, 220) write. The criterion used for distinguishing between biological inheritance and niche inheritance is direct versus indirect transmission rather than the three criteria I used (autonomy, near-decomposability, and content-dependent versus content-independent transmission). Space does not permit clarifying how my criteria relate to his. The reason I did not take direct/indirect as a criterion is that it rests on the common-sense assumption that biology is internal to the body while culture is not. Such an assumption needs extra justification. The defense of channelism given here is meant as such a justification and thus amounts to the same result: biological inheritance relies on biological reproduction and is distinct from the niche-cum-cultural inheritance.

It equally does not matter whether the biological and the cultural channel are further divided. Jablonka and Lamb (2005), for instance, distinguish between behavioral and symbolic cultural transmission systems and contrast these with two main biological transmission systems, the genetic and the epigenetic. Danchin et al. (2011) distinguish among genetic, epigenetic, parental, ecological, and cultural transmission. These authors, like Odling-Smee, thus use different partitions of the complete set of developmental resources, but they all draw an incisive and, from my point of view, decisive line between biological inheritance and cultural inheritance.

Genealogy and Channels of Inheritance 113

The divide between nature and culture is absent from, but nonetheless consistent with, how Griffiths and Gray (1994) partition different kinds of developmental resources (see figure 4.4). In their picture, biologically inherited developmental resources are a subset of parental resources. Cultural inheritance cross-cuts parental resources and collectively generated resources, the difference between the two being only that the first occurs in a vertical mode and the other in a nonvertical mode. One could thus easily superimpose the nature-culture divide on their partitioning.

The reason the divide is missing in Griffiths and Gray's (1994) picture is that, as they say themselves in the caption to the figure, "the effects of temporal order of interaction have been overlooked." For developmental systems theory, the primary aim is to understand development from a perspective that includes all the available developmental resources of organisms and to derive from that a picture of evolution that takes development seriously. Dynamic order, in particular the temporal order resulting from mostly vertical biological transmission versus changing transmission modes in culture, is unimportant since it is an evolutionary dynamic order that is irrelevant for their aim to take development seriously. I take this to be one reason for their initial holism with regard to inheritance, that is, their opposition to delineating different subsystems of inheritance.[15] To conclude, the temporal order between biological and cultural inheritance is developmentally irrelevant; evolutionarily, however, it is of utmost importance.

Result about Channelism, Genealogy, and the Explanatory Nature

As part of common sense, biological reproduction and social learning appear, first, as different kinds of interaction between organisms; they comprise different causal processes. The argument from autonomy states that the cultural inheritance channel is decoupled from biological inheritance (i.e., the transmission and thus distribution of cultural resources can change without a concomitant change in biologically transmitted developmental resources). This autonomy is backed by two further dynamic differences: the argument from near-decomposability shows that biological reproduction and social learning build channels (i.e., subsystems) of inheritance that are empirically discernible because they are near-decomposable; the argument from temporal order states that although both the biological and the cultural channel can be vertical and thus exhibit great evolutionary scope, cultural transmission is content dependent and because of that, quite particular in terms of stability and scope. Compared to the biological system, the stability and scope it confers can be much reduced. It is mainly (but not only) a fast track of evolution but therefore also often a less secure track.

For this chapter, the most important point is that genealogy grounds biological reproduction, which delimits the channel of biological inheritance, which provides a special stability necessary for homeostasis to be possible. The subset of the complete bundle of developmental resources of a species' life form that is particularly stable because it travels the biological channel can then legitimately be dubbed as the species' explanatory nature. Human nature in the explanatory sense is thus the pool of developmental resources that travel via biological inheritance. In chapters 7 and 8, this will be defended in more detail. In chapter 7, the stability that the biological channel confers will also be used to delineate a corresponding descriptive nature.

5.4 The Resulting Pluralism

From the frame that genealogy sets, three natures emerge that are distinct but connected via genealogy and provide answers not only to the partaking question but also to the description and the explanation questions. Yet the importance of genealogy is historically contingent. This shows that the analysis given here can only be a snapshot of our ideas about being human.

Three Natures Connected via Genealogy

A monistic account of kind's natures would assume that for natural kinds, there is one thing in the world that fulfills all (or more than one) epistemic roles simultaneously. The pluralist stance has three natures, each with its own epistemic role. The explanatory nature (i.e., the thing that replaces the explanatory essence) is a bundle of developmental resources. It is different from what replaces the definitional essence: a classificatory nature that is nothing but a relation (or, if one wants to formulate it as a property, the property of being genealogically related to other members of the species at issue), deciding the partaking question. The descriptive nature of a species consists of a property cluster—the typical properties that reoccur over time in a quite stable manner. Since the explanatory nature makes these properties widespread and stably appearing, it is causally involved in the production of the descriptive nature, but it is not the same as the descriptive nature and not the only thing explanatorily relevant for the evolution and ontogenetic development of the traits belonging to the descriptive nature.

The resulting pluralism is connected with a traditional ambiguity in the term *human nature*: traditionally, that term is simultaneously used to refer to an explanandum, the central traits of the human life form, and to an important subset of the explanans, the bundle of causal factors here taken to comprise the explanatory nature. This ambiguity has a long history. David Hume, for instance, used the term *human nature* in both ways. He looked for a science of human nature. Human nature was the explanandum, our typical and species-specific way of being (humanness), something that can be subject to law-like generalizations similar to the laws of nature. But the term *nature* was also used as an explanatory category—to refer to capacities that are "due to nature," that is, to describe what can—by common sense in Hume's time—be taken to be atemporally given as more or less inhering in more or less all people. In his famous account of causation, for instance, Hume argued that our causal inferences rely on habit, and habit was for him part of human nature (how humans naturally function). This meant that if one wants to understand why we talk and think about causation the way we do (an aspect of our descriptive nature), then one has to understand human nature (the habits of reasoning giving rise to it). Nevertheless, habits, which Hume took to be internal natural capacities for producing behavior, were not themselves Hume's target of explanation; they were taken as generalizations about humans that can safely be assumed as part of the explanans, as part of what explains human behavior, for instance, causal reasoning. Thus, the term *human nature* for him also referred to an explanatory nature that one can safely assume, but that one cannot—according to him—study any further, since hypotheses about its cause should according to Hume ([1739] 1896, xvii) "be rejected as presumptions and chimerical." Thus, habit is "a principle of human nature, which is universally acknowledged, and which is well known by its effects" (Hume [1777] 1902, 43). It is a "primitive element" of human understanding, as Norton (1993, 158) writes, interpreting Hume in the same vein as I do here. This pluralism about the term *human nature* is not overcome but used and radicalized in the account presented here.

The claim is that there is a classificatory nature, a descriptive nature, and an explanatory nature. For each traditional role of essences, there is something else that fulfills its role, but without being like traditional essences. One should distinguish the three different kinds of natures clearly, whether or not one agrees to use the analytic labels *classificatory*, *descriptive*, and *explanatory* and whether one agrees to use the term *nature* for them. Whether there are these three kinds of things that one can call a nature of a species is the factual issue; whether one should call them by the term *nature* is a terminological issue that will be dealt with in part III.

Genealogy's Relevance for the Explanatory and the Descriptive Natures
Given the pluralism just described, no one thing in the world answers all five species questions equally directly. Nevertheless, genealogy is ultimately involved in all the answers since genealogy is definitionally involved in the species and the partaking questions, directly explanatorily relevant for answering the constitution question, and indirectly and partly explanatorily relevant for the list of traits included in a respective field guide of the species, and thus for the explanation of the respective life form of a species.

The direct (often also called proximate) causal answer to the trait explanation question (and consequently the description question) is certainly the presence of developmental resources that cause an organism to exhibit the properties of the cluster. Yet the presence of the developmental resources is also in need of explanation. And the explanation of an especially stable subset of these resources involves genealogy: it is explained by the presence of the channel of inheritance that is characterized by high stability. Indirectly and partially, genealogy is then explanatorily relevant for the presence of an important set of developmental resources that allows for the life form of a species. The resulting genealogically grounded explanatory cluster of developmental resources is—because of its guaranteed stability and its explanatory relation to the descriptive nature—a good candidate for a species' nature in the explanatory sense.

Chapters 7 and 8 will deal with the stability that biological inheritance provides and other aspects of it in more detail since it is important to be precise about how to spell out this candidate for replacing an explanatory essence in order to not fall back behind the anti-essentialist and the interactionist consensus.

Historical Contingency of the Importance of Genealogy

There is a third, a historical and social value dimension of the pluralism defended here. If we were able to manufacture a population of *Escherichia coli* in the lab completely from scratch, so that there is no genealogical relationship of these laboratory individuals to natural *E. coli* individuals, would we call them *E. coli*? The point that I want to develop with this example is that it might well be the case that the more we would be able to do so, the more often we would call them so. Gradually our ontology would follow the lead of the technology: species would then become laboratory entities. The species category would become part of the laboratory's ontology and slowly cease to be mainly an evolutionary category. Right now, however, the evolutionary category is still influencing thinking about species, within and outside science.

The laboratory's ontology is a pragmatic context that clearly groups individuals according to their similarity and not according to their genealogical relatedness. In the language of Dumsday (2012), it groups individuals qua organisms, not qua species (given how we now use the latter term). Even though I reserved the term *species* for evolutionary units, the way that term is used outside evolutionary theory might change so much that there is no genealogical dimension of it left outside evolutionary theory itself (which will always be in need of an evolutionary group unit). For a time, the term *species* might be used in two senses (as I think it does already), as an evolutionary unit and as part of a nonevolutionary ontology, but ultimately it might completely lose its connection to evolutionary theory. The importance of genealogy might thus fade away.

This is so since concepts are creations of those having them and they change with the societies in which those having them are living. As Antony (2000, 19) wrote, "Definitional natures are created, either explicitly or implicitly, by human practices and conventions."

Furthermore, genealogy in and of itself (not how one gains knowledge about it) currently gets more and more complex because of new reproductive technologies (e.g., those used for surrogate motherhood). New social reproductive practices result from it. With the technologies and practices, the conceptions of reproduction, family, kinship, and partnership change too. Thus, I predict that the more complex the genealogical nexus between people gets, the more genealogy will lose its grip on our minds. Either a completely social concept of human nature (referring to a group delineated via social relations rather than biological relations) or a completely similarity-based approach to grouping humans will take its place. Nobody can tell what will happen, but something along these lines can happen. Some anti-enhancement activists already tell a frightening tale about the second option: we will live in a horrific world of similarity in which the homogenizing effect of using new reproductive technologies will be the price humanity has to pay for its liberties. Humans then would be regarded as laboratory animals, judged by similarity only (i.e., desirable traits), manufactured at will, mere objects of technological culture rather than people to whom one is related.

Let me add a historical remark about the word *kind* that makes vivid that an important concept, as tightly connected to genealogy and thus biological reproduction as the concept of human nature still is, can lose this vital connection. C. S. Lewis (1960, 24–74) in a study on the word *nature* and its cognates makes clear that the term *kind* originally referred to generation, sex, and biological reproduction and was related to *kin*, as in *kinship*, rather than to spatiotemporally unrestricted classes of things (as some historically unconcerned philosophers assume). Saying that a wo/man is kind was consequently pointing to a rather lascivious affair. The Indo-Germanic root of *kind* is believed to be *g'enə* (to give life) and the term *Kind* in German still denotes a child, as in "kindergarten." Given that the English term *kind* almost completely lost its connection to genealogy, the term *nature* can do so too.

This historical and social dependence of the importance of genealogy, however, threatens neither the realism nor the pluralism of the account defended here. Genealogy and the things that are meant by the three natures will still exist, whatever we do with our terminology. That there is a genealogical nexus between certain individuals that form an evolutionary unit will not change with the change of terminology.

That is why, from an ethical point of view, the pluralism defended here does not support an anything-goes attitude. There are matters of fact, for example, which developmental resources are present in an individual, equaling which explanatory nature an individual has. This then constrains which life form fits—irrespective of which label one chooses for the life forms. I thus argue that independent of potential future changes of the concept of human nature, it is a matter of fact that a child of two humans is an entity who shares what I call an explanatory and descriptive nature with other *H. sapiens*, whatever his or her parents believe with respect to how membership in a kind should be determined.

A couple in the United States tried to raise their child "species-less," by putting the child for months in a tree den so that the child could fully realize its tendency to self-identify with squirrels.[16] A reporter said in an interview with the child's father that "the facts are, scientifically speaking, that the child is a human being." The father replied: "Well, that's your label. That is a label we are putting out there for most life that is born to human parents and I don't think that is right." He is reporting that the label is conventionally used in one way but that he chooses to use it differently. He is certainly morally and legally free to choose his labels as he likes (at least as long as that itself is not imposing harm on the child). Yet he is not free to choose the way he treats his child. The child has, irrespective of the label and thus the group membership assigned, the same explanatory nature as other *H. sapiens* and is in a (nongenealogical) similarity sense the same organism as other *H. sapiens*. Similarly, a synthetic *E. coli* is the same organism as a natural *E. coli*, regardless of the historical difference of the two kinds and regardless of the labels one attaches to them. The child has probably paid a huge price for the parents not seeing this point. How one treats people should be decided by the facts: which life form is fitting to the developmental resources instantiated quite concretely in a particular child. Thus, based on the realism of my account, the parents should have been prevented from treating the child the way they did.

To sum up, a conceptual analysis has its limits. It can only be a snapshot of its time. When the technologies, societies, and values change, our concepts will too. Which classificatory concept of human nature one chooses involves moral and personal choices, not just scientific issues. Yet which genealogical nexus there is, which explanatory developmental resources an individual has, and which properties relate to it is not a matter of choice. The pluralist approach defended here is thus historical and pragmatic but not antirealist.

6 Toward a Descriptive Human Nature

In this chapter, I further defend the pluralism introduced in chapter 5 and address in which sense exactly there is a descriptive nature of humankind, a life form of the species *Homo sapiens*. I will claim that "human nature" in the descriptive sense refers to typical and stable properties of the genealogically identified humankind. That means, for descriptive purposes, typicality is a necessary but not sufficient criterion to decide which phenotypic traits are part of the descriptive nature of a species.

The issues around typicality have, like the issue about genealogy, an anchor in Hull's (1986) seminal critique of the concept of human nature. More recently, Machery (2008) defended a descriptive concept that he called "nomological." It relies on typicality and evolvedness to decide what belongs to human nature in the descriptive sense. Hull's and Machery's accounts will be the starting point for a new solution, more pluralist than both, with a focus on the issue of abstraction, to be discussed in this chapter, and with a replacement for evolvedness as the second criterion, to be discussed in chapter 7.

After discussing the need for descriptive knowledge about humans in general in section 6.1, I elaborate on these relationship between descriptive, classificatory, and explanatory knowledge in section 6.2 to bring to the fore the advantages of the pluralism defended here. In section 6.3, I discuss whether typicality is necessary, given critiques of this criterion, mainly with respect to polymorphisms. That typicality is not sufficient will lead back in section 6.4 to a claim that stems from David Hull: traits that are part of human nature are somehow important. In which sense they are will lead to stability as the second necessary criterion for specifying what is part of the descriptive nature of humans.

6.1 Descriptive Knowledge about Humans in General

When Hull laid the foundation for a genealogically specified classificatory concept of human nature, he did not deny that there is character covariation—a so-called phenetic cluster, a cluster of typical properties, be these physical, behavioral, or mental. The cluster of properties provides generalizations that allow reliable inferences—predictions from one property's presence to another's, from one kind member to another kind member. Taken together, the generalizations allow a description of the life form of a species and can thus be regarded as a descriptive human nature.

Hull regarded any such cluster of typical properties as indecisive for membership of individuals in a species. I agree with him up to this point, but character covariation is more important than Hull admitted since the concept of the nature of a species can be used for different things. Classification of individuals as partaking in a species is not its only epistemic role, as the distinction of the five questions in chapter 5 illustrated. Description and explanation have repeatedly been mentioned as equally important epistemic roles.

Actually, even Hull accepted a version of a descriptive nature as existing, for example, when he wrote:

If by "human nature" all one means is a trait which happens to be prevalent and important for the moment, then human nature surely exists. (Hull 1986, 9)

Prevalent is Hull's word for what I call *typical*. There clearly are historically changing clusters of typical properties. The respective properties are—at one point in time—typical (and some even specific) for *H. sapiens*. The term *typical* sometimes implies "ideal" representation, but in a post-essentialist frame, the term is used in a more minimal way, merely pointing to distribution and frequency. To be typical, a trait needs to be species-widely-distributed and frequent. (I will defend both these qualifiers, "species-widely-distributed" as well as "frequent" below.)

Such a descriptive nature exists despite two important issues. First, it exists, even though the respective properties are not necessary or sufficient conditions for individuals partaking in the species (since not all and only humans exhibit them) and, second, although knowledge about which properties are part of the respective descriptive nature is hard to establish. The first issue is solved since a pluralist picture decouples the descriptive from the classificatory role. I will say a bit more on this in section 6.2. The second is an issue still often overlooked but of quite some importance. As Henrich, Heine, and Norenzayan (2010) stressed vividly, generalizations about humans need to be made very carefully; unfortunately, they rarely are. When data are collected about humans, for instance, in psychology most of the time "WEIRD" people (for Westernized and educated people, from industrial, rich, and developed countries) are chosen as representative for all humans. Henrich et al. show that these are often far from representative—be it with respect to visual perception, fairness, cooperation, spatial reasoning, categorization and inferential induction, moral reasoning, reasoning styles, self-concepts, and related motivations, or the heritability of IQ. WEIRD people are "frequent outliers," or simply "the weirdest people in the world" (now in the conventional sense of the word). The conclusion is that "we need to be less cavalier in addressing questions of human nature on the basis of data drawn from this particularly thin, and rather unusual, slice of humanity" (Henrich et al. 2010, 1). Although they stress the need for higher standards in establishing generalizations about humans, they do not question that there can be such generalizations. In other words, they do not question that there is a human nature in the descriptive sense. Thus, although there is an important methodological issue, it is not threatening the adequacy of the concept of a descriptive nature.[1]

The problem I have with the minimal concept Hull left alive is that it ignores that a nature of a species usually not only applies to all contemporary humans but also to past humans. The sharing of human nature extends to past generations too. The descriptive nature has a diachronic dimension since individual members of *H. sapiens* resemble each other synchronically as well as diachronically, and usually in many more than one character, although they are not the same (share all the characters). The cluster of properties making up human nature in the descriptive sense is, metaphorically depicted, not like a set of circles (as in the figures in chapter 1) but more like a set of clouds, extending in space and time. Properties that are part of human nature reliably reoccur over time, not just across space. Hull seems to have kicked out that diachronic dimension (and many post-essentialists followed him in this), but as I shall argue, he only hides the temporal dimension in the qualifier "important," which is the qualifier he inserted in his watered-down concept of human nature, but without explaining what it means.

Given that human nature in the descriptive sense extends in space and time, one could either take "typical" to refer to species-widely-distributed properties that are shared by most individuals at a certain moment and add stability of prevalence over time as an extra criterion; or one could include the stability over time as a temporal dimension of typicality since this is basically what it is. Not much depends on this terminological choice, but since the majority of people use "typical" without a time dimension, I will do so too and rather defend this characterization:

> Descriptive human nature = traits that are typical and stable for the genealogically identified humankind.

The stability of the property cluster is what often makes some typical properties important. I will add reasons, but the punchline can be clearly stated now: the claim is that stability is needed as an extra criterion since typicality is not sufficient to narrow the list of traits that are conventionally accepted to belong to human nature. We need the temporal stability to not end up with a concept that is too broad. But before, the need for a descriptive human nature needs to be defended.

The stable property cluster, the descriptive nature, is what Hume and others in the Enlightenment were referring to when they were talking about a "science of man," a science *of* human nature. The descriptive nature is thus, first and for all, a unifying explanandum of all those concerned with humans in general. In addition, some traits are, as Machery (2008) and Griffiths (2011) mention, indispensable for certain sciences, such as anatomy, physiology, paleontology, archaeology, molecular anthropology, and cognitive-behavioral research fields such as ethology or psychology. They can function as specific explananda or as part of an explanans. A trait like the human hand (with its opposable thumb) is, for instance, relevant for social scientists who explain tool use, the evolution of mind, and culture. Ayala (2016, 44) points at that when he writes, "Hands allowed our ancestors to make tools, which in turn led to an enlargement of the brain and finally to the enormous intelligence that we have. Due to the enormous increase in intelligence, a new kind of distinctively human evolution appeared: cultural evolution—cultural adaptation to the environment." This might well be wrong empirically, but it shows how general knowledge about the human hand (or whatever other part of our body that is relevant for cognitive and social differences) is relevant. Even if it is not itself the explanandum, it can still play an explanatory role, as part of this or that explanans. Having a language and having intentionality are equally uncontroversial parts of human nature that are of highest importance: as explanandum, as, for instance, in psychology; or as part of the explanans, to study something else, say, history. If the latter is the case, as in many fields in the humanities or the social sciences that deal with concrete social problems, such as development studies, human nature (in the descriptive sense) might even be treated as a disciplinary primitive, something assumed as part of the explanans but not studied. It is what is shared among humans, the foundation without which the diversity of phenomena studied by the humanities would be impossible. Thus, even the humanities and some social sciences (such as cultural anthropology or social psychology, which are interested in cultural differences) need a concept of descriptive human nature, that is, reliable generalizations about humans, even though often as a mere contrast foil or background condition. As hinted at above, anybody doing historical research relies on the assumption that the subjects studied are humans capable of producing human language (otherwise one would be doing ethnology or paleoanthropology rather than history). If the descriptive human nature is a disciplinary primitive, it is assumed or left in the background in the sense that humans in general are not studied, but specific cultural or historical traits that are (in the language of biology) polymorphic. Thus, fields like gender studies, religious studies, history, and sociology all study polymorphic traits, particularities related to the "things" produced by specific groups of humans: gender stereotypes, Australian religions, Chinese history, African literature, European art, and so on. Humanities do not study human nature; they study the reliably occurring consequences of human nature, such as history, art, institutions, and texts—in short, culture. Finally, the example of the parents aiming to raise their child species-less, which I briefly discussed in the previous chapter, signals that the descriptive nature is also of ethical importance.

Neither the Darwinian challenge nor the developmentalist challenge contests the status of such a descriptive human nature. Descriptive human nature exists. The search for human nature in the descriptive sense simply is, as Dupré (2011, 169–170) puts it, a "search for the best characterization available of what humans are like."[2]

I first describe in more detail in section 6.2 in which sense a descriptive human nature is decoupled from the classificatory and the explanatory role. In section 6.3 I justify why typicality is necessary for a trait belonging to the descriptive nature, while in section 6.4 I argue that it is not sufficient since stability over time is needed too. By looking at typicality as necessary but not sufficient, one can make progress in ordering the diversity of synonyms used for "typicality" (such as *characteristic, universally distributed, prevalent, shared by most humans*), but also in clarifying the mysterious qualifier in Hull's phrase—that traits that belong to human nature are somehow important.

6.2 The Relationship to the Classificatory and the Explanatory Nature

Decoupling of Descriptive and Classificatory Role: Variation Becomes Bearable

No classificatory role is attached to the descriptive human nature. The two roles are decoupled. As a consequence the term *typical* is not to be treated as synonymous with the term *universal*. If human nature in the descriptive sense is conceived in a minimalist (i.e., not claiming more than needed, given a specific epistemic goal) and pluralist manner (i.e., decoupled from other epistemic roles), it does not matter that none of the qualitative, typical, and stable characters are necessary, nor is any combination of them sufficient for being *H. sapiens*, shared by strictly all and only humans. What matters is that there is a historically slowly changing stable cluster of statistically typical properties of humans.

Such a cluster, however, is not a property of an individual; it is rather a property of a population, since none of the individuals exhibit the cluster. It is a statistical pattern, allowing for statistical generalizations. A property cluster (a property of a population) should thus not be confused with a cluster of individuals or cluster of properties within individuals.[3] Together with the issues discussed in chapter 3, this suffices to derive that knowledge of a property cluster will not provide an answer to the question of whether a specific individual belongs to the group of individuals correlated with the cluster. Since the property cluster is statistical and therefore populational, one cannot infer from it that an individual counts as a member of the population. This is not a matter of myopic stance, but a matter of principle: stemming from the facts about variation, from the fact that it is about an empirically established cluster (rather than necessary and sufficient conditions) and, finally, that it is a cluster of properties (rather than individuals) not necessarily intrinsic to individuals.

Luckily, that is all one needs in order to talk in meaningful descriptive manners about humans in general. Variation, which was a major problem for the essentialist, becomes bearable by decoupling the descriptive and the classificatory nature and by going populational.

The Decoupling Allows a Noncircular Way to Determine the Reference Class

I assumed in chapters 3 and 5 that one decides who counts for certain generalizations by having membership conditions that answer the partaking question. The so-called reference class, the extension of statistical generalizations, is then fixed by these membership conditions.

For instance, if the classificatory nature is the genealogical nexus, then the reference class of the descriptive nature (the statistical generalizations about humans in general) is fixed by genealogy. To recall the results from chapter 5, in all situations where there clearly is a species *H. sapiens,* the genealogical nexus is necessary and sufficient to answer the partaking question of which organism belongs to the respective species. In all other cases, the question does not make sense. As described in chapter 5, to choose the genealogical nexus as necessary and sufficient for membership in the kind is a pragmatic decision that depends on scientific usefulness, social experiences, and values and these might well change over time. Yet whatever change there is to how the classificatory nature is determined, there has to be some fixing of the reference class independent of the descriptive nature.

In theoretical accounts of the concept of human nature, and also in practice, that first step of fixing the reference class is rarely made explicit.[4] If one were to answer—in a nonpluralist manner—that the reference class is decided as a result of observing generalizations, regarding phenotypic properties, then the result would be a classificatory circle: one would have to pick out some trait (or traits) as relevant for classification, according to which one establishes the members of the group in which one looks for clusters, and then one establishes the generalizations about the same traits of that group. This is circular. Furthermore, depending on what one picks as decisive for membership in the reference class (and be it one trait or a set of traits), one will pick out a different reference class, as illustrated in chapter 3 with respect to the problem of squaring the circles. Consequently, one will not only pick a different property cluster but also a different set of individuals. The core of the problem is that most of the time, one will find a cluster, whatever one picks as a property to start with. If the objection is that there are clusters of properties in the world, then the answer is: certainly, but there are too many and there is no simple hierarchy among those nongenealogical properties that are driving the clustering. As Hull put it:

> If the history of phenetic taxonomy has shown anything, it is that organisms can be subdivided into species as operational taxonomic units in indefinitely many ways if all one looks at is character covariation. (Hull 1986, 11)

Thus, one would equally end up with a pragmatic choice if one picked a qualitative trait (or set of traits) to first fix the reference class over which one then makes generalizations. In addition, the method would be rather circular and end up in a pluralism that is too promiscuous. Finally, if one were honest, one would simply admit that with respect to humans, the reference class is evident, regarded as fixed from the very start. As already mentioned in chapter 5, I assume that at least the majority of people today still assume that all and only those who are genealogically related to other *H. sapiens* are *H. sapiens*. As Antony (2000, 18) remarks, generalizations in those sciences concerned with humans (including humanities) with respect to humans are about "individuals *antecedently* identified as humans [emphasis added]." The identification of members of the species is logically and procedurally prior to generating the generalizations for the description of the life form of the species. The property cluster depends on the reference class but not vice versa. The relationship between the membership conditions and the property cluster is asymmetric.

Certainly, and as discussed in chapter 5, the defining and the identifying criteria can diverge in the prior fixing of the reference class. In fact, they do so in the case of defining *H. sapiens* (genealogical criterion) and identifying *H. sapiens* (phenetic criteria). Genealogy is the logical criterion one uses for classification of individuals as humans, even though one rarely checks for it explicitly and rather uses reliable indicators (i.e., epistemic identifiers). Genealogy is not directly observable since if one sees a person, one does not see the person's genealogy. Genealogy is assumed or inferred or testified (e.g., with a birth certificate). That one does not check for the actual defining criterion (genealogy) adds plausibility to the argument that there is usually no practical problem of deciding who belongs to the species *H. sapiens* (except for the beginning of *H. sapiens*).

Let me sum up the decoupled picture of the relationship between the classificatory and the descriptive nature. First, there is a bare numerical identity that I call classificatory nature (call it classificatory essence, if you like) that determines *who* P is (not *how* P is). It decides species membership (or partaking in a species, to stay metaphysically neutral). It consists in the genealogical relation to other members of the species. It determines the position of an individual on a chunk of the tree of life; it determines the belonging of that individual. Second, there is also the life form of a species, a descriptive nature. It describes *how* members of the species are. It provides a qualitative identity.[5] It consists of a stable property cluster that extends in space and time and holds for an antecedently fixed reference class. The next step is to explicate the relationship that the descriptive nature has with an alleged explanatory nature.

Relationship to the Explanatory Nature: The Descriptive Nature as an Etiological Category

From Sober's arguments in chapter 3 it follows that the descriptive nature of a species is not explained in a traditional essentialist sense with reference to a natural state with more or less good realizations of such a natural state. In the language of Machery (2008, 323), the descriptive concept "inverts the Aristotelian relation between nature and generalization," a relation that is, according to Sober (1980), inscribed in the natural state model. If there were such an Aristotelian relation, it would be the case that having "the same [explanatory] nature *explains why* many generalizations can be made" about the members of a species, despite variation (Machery 2008, 323, emphasis added). For Machery, who argues for a descriptive concept of human nature that he calls "nomological," it is simply the other way around: "the fact that many generalizations can be made about humans *explains in which sense* there is a human nature" (emphasis added), namely, as I would say, in a descriptive sense.

I agree that this inversion of the Aristotelian relation between nature and generalizations is of utmost importance to not fall back behind the dehumanization and the Darwinian challenge. The only problem that I have with Machery's way of putting things is that the term *explains* changes its meaning in his inversion. In the inverted form "explains" is not referring to any causal explanation but to explication. The inversion means that if reliable generalizations about humans are possible, one understands what it means to have a nature in the descriptive sense. That's it. Yet I agree with Machery's main philosophical point regarding his "Aristotelian inversion," which I take to be that having a descriptive nature is not explaining why scientists can make the reliable generalizations about humans that they in fact make. The descriptive nature is nothing more than a bundle of generalizations about humans. If used in explanatory contexts, it is the explanandum rather than (part of) the explanans.

In a new approach, however, Machery (2016a) develops an account in which the descriptive nature itself nonetheless carries explanatory information. To arrive at this, he shows that human nature is a so-called etiological kind term, referring to a special kind of kinds—one in which membership in the kind depends on the etiology (i.e., causal history) of the entities grouped into that kind. If the question is which traits (of humans) belong to the set of traits (i.e., the kind of properties) included in human nature, then only traits with a specific causal history belong to that set of traits. According to Machery, the respective causal history must be an evolutionary history. Thus, only typical and evolved traits ("evolved" taken to refer to a causal history) are part of the etiological kind of traits that are part of human nature.

The explanatory power of such an etiological kind term is then the following: by grouping something into such an etiological kind (e.g., a trait as being part of human nature), one gets, just by that epistemic act, the explanatory information that the trait has a specific kind of etiology (i.e., an evolutionary one). If one gets told that bipedalism is part of human nature, then by that alone, one can infer that it has an evolutionary explanation. "Classification in etiological kinds is causal explanation without causes"; it is giving an "explanatory sketch" (Machery 2016a, 220). Yet an explanatory sketch would still only be a shortcut, an abstraction from the actual causal explanation—those explanations that do refer to causes. The difficult question thus remains: What is actually taking over the role of the explanans? The descriptive human nature does not fall down from heaven like manna. I side here with Samuels (2012), who is the direct opponent Machery (2016a) addresses when he introduces his view about etiological kinds. At issue in their debate is which nonessentialist successor concept, the nomological one or an explanatory one Samuels suggested, is better to fulfill both the descriptive and the explanatory roles. Whereas they both still argue in a monist spirit (i.e., they look for *the* successor concept), I opt for a pluralist solution: the nomological concept refers to traits and better fulfills the descriptive role and there is a separate explanatory human nature concept that refers to causes (not just explanatory sketches) and better fulfills the explanatory role. I agree with Machery that one cannot have both at the same time (as Samuels seems to want it in his account), but I also agree with Samuels that the explanatory sketch is inferior to a full-blown explanatory nature.

In chapter 8, I argue in which sense developmental resources of a specific sort, namely, those handed down via biological inheritance, can be counted as explanatory nature. In the remainder of this chapter, I justify that typicality is necessary for a trait to belong to the set of properties that make up the descriptive nature of humans and that it is not sufficient since stability needs to be there too.

6.3 Typicality Necessary?

Is typicality (in the sense of species-widely-distributed, frequent properties) really necessary for human nature? There are two ways that typicality can be questioned.

Species Specificity Rather Than Typicality

Some philosophers might argue that the only thing that is necessary for human nature in the descriptive sense is species specificity (i.e., uniqueness). I argue that species specificity is reducible to species typicality if it is taken to refer to a descriptive nature rather than to a classificatory nature.

At first glance, specificity seems to relate to the classificatory role of the traditional concept of human nature, a role that can and should be decoupled from the descriptive role (see above). The function of reference to species-specific traits is often to demarcate one kind of animal from another kind of animal. Given the Darwinian challenge, that function collapses. There are no strictly species-specific human traits that no individual of another species can have. Yet one can treat specificity in a statistical and nonclassificatory manner, but then we reduce it to a special kind of typicality: what is typical in one kind only. After all, what is specific for humans should also be typical for humans, at least if that species-specific trait should count as part of human nature in the descriptive sense.[6]

I conclude that if we reduce specificity to typicality in an exclusive sense—what is typical in one kind only—then one is replacing typicality with a specific sort of typicality rather than challenging that it is a necessary criterion for something to be part of human nature.

Furthermore, to claim that exclusive typicality (typical of one kind) delineates which properties are part of a species' nature might suit the purposes of the philosopher, interested in differences between humans and other animals, but it will likely not suit the epistemic interests of the biologically minded: having a heart and a lung is, after all, part of human nature, even if it is not part of our nature exclusively. There are many typical traits that are not species specific and quite important for understanding how humans are, even if they might be quite unimportant for the aims of the philosopher, who often has interests of her or his own. Thus, what is important (in Hull's sense mentioned above) varies with disciplinary affiliation or interests. Emotional facial expression, for instance, seems to be quite typical but not specific for humans, as Darwin claimed.[7] And while emotional expressions have rarely been at focus in philosophical theories about human nature, they are quite important in other fields. I will come back to this inner-scientific perspectivity.

In any case, this brief discussion of species specificity suffices to confirm that typicality seems to be often invoked (in the philosophically minded and the biologically minded) as a necessary criterion for something being part of human nature, in an exclusive or a nonexclusive sense.

Polymorphism

A second way to challenge typicality as a necessary condition for a trait counting as part of the descriptive human nature is to point to polymorphisms: traits that are regularly expressed in one and the same species in many (hence, poly-) forms (hence, -morphic). Blood type is a polymorphism. So are sex-specific traits. The properties that derive from the sex of a human individual (e.g., the property "being able to give birth to a child") are exhibited by roughly only half of all humans. It would be a stretch to call sex-specific properties of humans typical for humans. They are typical for women or men but not typical for all members of the species since the respective traits are not frequent (most of the members have it), even though they are widespread (found throughout the species).

In principle, one can still consider such polymorphic traits as part of the respective species' nature, despite the fact that it is regularly not the case that the majority of individuals express them. A species' life form would then be understood to consist not only of similarities but also of stable and widespread patterns of difference, since that is, after all, what polymorphisms are. Species members are similar in stable ways, but they also differ in stable ways.

To stick to typicality as necessary and do so without ignoring polymorphisms, one can include polymorphic traits as disjunctive traits, as Lewens (2012) suggested and Ramsey (2013) defended. As a result, human beings are either female and have property X1 or male and have property X2 (which can be the negation of X1). Hull (1986) mentioned such a disjunctive solution with respect to blood types in humans.

Machery (2008, 2012), by contrast, excludes polymorphic traits from being part of human nature by assuming that only those traits should be regarded as part of human nature that are "widely shared" (i.e., frequent). We might regard sex-specific traits as part of the respective nature of human females or the nature of human males, but they are, according to Machery, not part of human nature. By doing so, Machery (2012, 477) assumes that traits that are species-widely-distributed, that is "found throughout the species," are included only because being widespread is a proxy for being "widely shared" (frequent) properties. The problem is that polymorphisms are an exception. They are species-widely-*distributed* (i.e., widespread) but not *frequent* (i.e., widely shared), and thus not typical for all humans. On the basis of this, Machery argues against including, for instance, sex-specific properties as disjunctive properties of human nature.

> It would be arbitrary to include the traits of morphs that are present throughout a species (e.g., males and females) into its nature, but not the traits that are distinctive of lineages within that species, such as subspecies that have evolved in different environments. (Machery 2012, 477)

I would like to drive this point to its extreme. In the limit, any pattern of variation in principle can be included as a disjunctive property. The disjunctive solution is thus in danger of making the concept of human nature too inclusive. For any property (e.g., having five fingers), there can be an open list of alternative properties (e.g., one, two, three, no fingers, something like fingers) so that an abstracted property (e.g., having some or something like fingers) is constructed out of the pattern of disjunctive differences. Imagine that a set of new tribes of humans is discovered, never seen before, and the discoverers realize that for this population, the trait of "having five fingers" is not typical. Then more such strange groups are discovered. They conclude that humans have "five fingers or four fingers or three fingers, or two things like fingers." They conclude that there is a quite complex pattern of similarity and difference with regard to what is called "a hand." Hence, they decide to replace the trait "having a hand with five fingers" with the trait "having a hand with some or something like fingers," a trait abstracted from the pattern of diversity of actual hands. That way, with such an abstraction method from disjunction, everything can epistemically be made part of human nature by simply abstracting away from the differences actually observed. Thus, including polymorphisms the way Lewens (2012) and Ramsey (2013) suggest is leading into an infinite regress: everything can be included as part of human nature. I call this the *disjunctive regress argument*. I regard it as a strong argument against including polymorphic traits as part of a species' nature.[8]

Yet, there are two issues that need to be discussed with respect to it. First, Ramsey (2013, forthcoming) defends an account that bites the bullet: he defends a concept of human nature that indeed includes every possible human trait, regardless of its prevalence. Thus, typicality for him is not necessary for a trait to count as part of the descriptive nature. Second, the disjunctive regress goes in two directions since uncontroversial typical traits can also be described as disjunctive traits. This will illustrate why the pluralism defended here is so important.

Ramsey's All-Inclusive Account of Human Nature

Ramsey (2013, forthcoming) treats every trait that a human can possibly develop as part of human nature. Even PKU, a rare disease, is part of what he suggests regarding as human nature simply because it is possible for humans to get the disease; it is within the constraints of what the current pool of developmental resources allows.[9]

The problem with such a broad notion of human nature is not that it is far from the standard way of using the term *human nature* in sciences since (as I will show below) there is no such standard way; the problem is that it lacks contrastive power. It is too inclusive. With contrastive power, predictive power is diminished too. Machery (2016a) pointed at the latter, reacting to Ramsey (2013): if every phenotype "that a human being could have is included within human nature, ... one cannot justifiably infer that a human being is likely to possess a trait from the fact that this trait belongs to human nature." I take Machery's point to be that to infer that something is likely since it is possible is not a very reliable inference pattern. Ramsey (forthcoming, 54) replied that this "misses the point" since "when we make predictions about which traits an individual will bear, we usually do not do so in the total absence of knowledge about them. Instead, we make predictions based on what other life history traits they possess." This is true, but indeed a case of talking past each other. Machery's point, it seems to me, is that a proper concept of human nature has predictive power even in the absence of knowledge about further traits of particular individuals. Just by knowing that a trait is part of human nature (in the absence of any further information except that the individual is a member of humankind), you can predict that it is likely that the individual will have the trait at issue. That is not the case for PKU or any of the other nonprevalent properties that Ramsey counts as part of human nature, even if they have a nonrandom pattern of occurrence, like PKU (and many other traits). Just by knowing that an individual is a human, you can predict that it is likely (even if not certain) that the individual will walk upright and speak, but you cannot predict that it is likely that she or he has PKU; on the contrary, it is rather unlikely that he or she has this rare disease. And this holds despite PKU's "robustness" (Ramsey's term for flatness of norms of reactions), which PKU undoubtedly has (except for the situation where a phenylalanine-free diet is kept). One can predict the disease very reliably: if the respective mutations in PAH occur (and the diet is not possible), it is almost certain to occur. This means that PKU is much determined by biologically inherited developmental resources and is due to nature in that sense. One can then call it part of an *individual's nature*. Yet we should not confuse that context of using the term *nature* with the context in which the term *nature* refers to a species. A species' nature is not just all individual natures put in a summative bin, but that is exactly what Ramsey suggests.

In addition, the following three points apply. First, an all-inclusive human nature will result in a situation where the claim that human nature exists is tantamount to the claim that it does not exist. A concept of human nature that includes every possible trait is empirically indistinguishable from the claim that we have no such nature, that we are free in the sense that existentialists like Simone de Beauvoir and Jean-Paul Sartre stressed, a sense that implies limitless freedom. The empirical pattern assumed in the account I defend is, by contrast, the following: human nature changes over time; there are not just commonalities but also differences between humans, and over the long run, we can influence what is common (i.e., typical) and what is varying. Depending on the patterns of similarities and differences, that is, whether humans are empirically a species with a lot of commonalities and a low degree of differences, the descriptive human nature is thick or thin. Every species has a descriptive nature, but whether the descriptive nature is rather thick or rather thin depends on the case and can only be established empirically and comparatively. And from all knowledge science has produced about humans, it is pretty clear that the descriptive nature of humans is quite thin compared to other species, thanks to our evolved freedom, but still quite thick compared to a random assortment of individuals from a diversity of species. Neither the existentialists nor Ramsey allows us to distinguish between a thick or a thin descriptive nature. The result is that predictive power is almost completely lost.

To conclude, an all-inclusive concept of human nature is useless for predicting which properties to expect once we know (and nothing else) that an individual is a member of the species. That predictive power is important since we are practically often in situations where we actually lack the knowledge that Ramsey claims we often have. When simply meeting a person for the first time, be it in rescuing a person or otherwise interacting with one, one needs to rely on knowledge about humans in general, for example in order to decide which actions are adequate. The same holds for sciences, which use the sheer fact that somebody is human to make social psychology experiments manageable and precise, without checking every aspect of all the study participants.

Ramsey (2013, 989) seems to try to reintroduce such a predictive power by distinguishing between core properties of human nature and less core properties. The suggestion has not been fully developed, but it seems to treat a high p-value (p for prevalence, hear called typicality) as able to "restrict" what counts as core. This would reintroduce what has been kicked out before, and I do not see what is gained by Ramsey's account if it in the end reasserts that typicality is core for a concept of human nature. He even mentions, even if only briefly and with different intentions and probably meaning something slightly different, that "one can achieve a high p by simplifying" the way the trait (or trait associations in his case) is represented. This simplifying is what I call abstraction.

Typical Traits Are also Disjunctive

I have described how polymorphisms or any trait can be made typical by abstracting away from the actual differences and inventing a new name for the abstract similarities behind the differences. This was the basis for claiming that including polymorphisms into human nature is possible but ultimately leads to an infinite regress since then everything about humans can be included as part of human nature.

The method of abstracting away from differences points to an even deeper problem since the regress goes both ways. If we exclude polymorphisms since they are disjunctive, we also have to exclude some traits that are standardly taken as part of human nature, such as the ability of humans to have a language. After all, that is also a property that is abstracted from the diversity of concrete languages that concrete human beings exhibit. Each human has this or that language and not language *simpliciter*. But "having a language" is regarded, presumably even by Machery, as a proper general trait of humans, whereas polymorphisms are contested. What is the difference? It cannot be inclusion by disjunction and abstraction since both are actually arrived at via disjunction and abstraction.

The core of the problem is that there is neither an a priori argument on where abstraction has to stop nor any general empirical rule where to stop. There is no "natural level of abstraction" as Antony (2000, 35) mentioned with respect to how difficult it is to empirically decide whether a trait in question is truly typical for all humans. In other words, if there is no rule for when abstraction from differences has to stop, then the regress can spread in both ways: a trait such as language is as much a victim to the disjunctive regress argument as a polymorphic trait. Human nature is then simultaneously in danger of being too thin (regress uncovering typical traits as being abstractions from disjunctions) and too thick (regress spreading toward all differences being included in human nature).

Any actual abstraction stopping will involve pragmatic decisions. An answer to how and why scientists stop abstraction for some traits but not others would require a detailed look at the respective scientific practice. In the case of biologists, this would involve understanding, first, how biologists actually distinguish polymorphisms from other kinds of variations, and, second, how the abstraction of properties is regulated in context, that is, which abstractions are legitimate and which are spurious. Without an empirical analysis of the practices in specific scientific fields, an analysis that cannot be given here, a general discussion of this leads nowhere. The debate between Lewens, Ramsey, and Machery cannot have an a priori general philosophical solution.[10]

I derive two general points from the philosophical analysis given above, points that are independent of the results of an empirical analysis of the practice of abstraction. First, to opt for inclusion or exclusion of polymorphic traits is a choice that does not affect the meaning of the concept of human nature. Irrespective of whether one chooses to include polymorphic traits (as disjunctive trait or as a general trait abstracted from the disjunctive trait) or to exclude them by definition from the list of traits belonging to human nature, the term *human nature* would in both cases refer to stable and typical (i.e., widespread and frequent) properties of the life form of a species rather than to properties of a subpopulation, which would be called the nature of that subpopulation. Machery and Lewens at least agree that the reference class of human nature stays the same: it is the species rather than a subpopulation. They thus agree that typicality for the species is the decisive criterion for including traits as part of human nature. They disagree whether adding disjunctive traits is advisable and maybe where to stop the practice of abstracting from the pattern of diversity (that comes with any biological trait) in order to arrive at a typical property. From a pluralist point of view, the decisions on whether to include polymorphisms or not are pragmatic matters, determined by the context of research. If a subpopulation is an interesting group, it is likely to get attributed a separate nature; if the differences in the subpopulations are not of interest, the diversity gets bundled under a general property. As an example, evolutionary psychologists sometimes care to explain sex-specific cognitive properties, whereas a taxonomist might well be completely uninterested in that pattern of differences. The former is likely to speak about the nature of human females (to stress sex differences), while the latter is not. I take Machery (2008, 2012) to agree on this since he admits that it might be a pragmatic issue whether polymorphisms are included in human nature or not.

Second, siding with Machery against Ramsey, I claim that in many contexts, predictive power should be an important part of the evaluation of where to stop abstraction from differences. And for the predictive power, typicality is necessary. An all-inclusive concept has contrastive power only for one thing: distinguishing those sciences studying humans (human sciences) from those that do not study them at all but rather look at the universe, rocks, or whatever else. It does not help to understand the different epistemic interests, such as studying similarities (e.g., as in cognitive psychology) and studying differences among humans (e.g., as in social psychology or so-called cognitive difference research).

Machery, Lewens, and Ramsey all more or less accept that pragmatics enters the business, but they all argue too monistically, looking at only one epistemic interest: describing the overall patterns of similarity and difference in the case of Ramsey; describing the pattern of species-wide similarities contrasted with describing species-wide differences in the case of Machery and Lewens; Ramsey ignores that sciences are by now specialized into describing similarities or differences, and Lewens and Machery ignore that there is no hierarchy between describing similarities and describing differences. A pluralist picture, as defended here, regards the different contexts as symmetric. I conclude that it depends on the context of research whether the abstraction of a typical property (from the disjunction of the observable differences) is adequate. The details of the actual research matter, but not for the necessity of typicality for a descriptive concept of human nature that has predictive power.

6.4 Typicality Sufficient? Or What Does "Important" Mean?

Our recently acquired habit of constantly carrying cell phones is quite typical and occurs across generations. It extends not just in space but also in time and can be expected to exhibit some stability. Yet it is a habit that is learned rather than biologically inherited. Would we call this habit a part of human nature? It depends on the context.

In certain scientific contexts (descriptive contexts in the human, cognitive, or social sciences), some might well say yes, that is part of how humans are now, and in that sense it is part of human nature. I consider the concept of human nature defended by Stotz (2010), Griffiths (2011), Stotz and Griffiths (forthcoming), and Ramsey (2013, forthcoming) to be committed to such an answer. It treats typicality as a sufficient criterion to regard a trait exhibited by humans as part of the descriptive human nature. (As described, this does not entail that they regard typicality as necessary.)

If every typical trait is regarded as part of human nature, then classifying the trait as being part of human nature does not provide much explanatory information. Treating the reliably reoccurring cell-phone-carrying habit as part of human nature might be adequate in certain epistemic contexts, but in others, it fails since traits that are typical but clearly socially learned rather than biologically inherited are explained differently, despite the interactionist consensus. Taking this difference into account adds explanatory power. The cell-phone-carrying habit and bipedalism belong to different etiological kinds, as Machery (2016a) already stressed. In contrast to Griffiths, Stotz, and Ramsey, I take Machery (2008, 2016a, forthcoming) and Samuels (2012) to be committed to such a more nuanced account, which in the end involves the nature-nurture divide in some way or another–for the sake of explanatory power. What is wrong with Machery's and Samuels's account is, as I will claim below, not that they use the nature-nurture divide but how they use it. I will develop a version of it that is compatible with the developmentalist challenge. The solution to the developmentalist challenge is not a broad concept of human nature that throws out any distinction between human nature, culture, and environment.[11]

From what I have said so far, I derive that in certain contexts, typicality seems to be insufficient to narrow down the list of traits so that the concept of human nature (as an etiological kind) has the predictive and explanatory power that is desirable in many contexts. Some people might nonetheless simply not care about prediction and explanation. It is legitimate to simply care about describing typical traits if one is interested in pointing out similarities among humans (e.g., in the humanities), without being wedded to a certain kind of explanation—and that entails, as I will show, a certain kind of stability.

But those who claim—as Machery, Samuels, and I do—that typicality is in many contexts not sufficient might well disagree on how to narrow down the list of traits. In addition, the philosophy of science literature on human nature so far suffers from a too narrow focus on certain disciplines. Machery cares for evolutionary-explanatory business, for "what evolutionary behavioral scientists typically mean by 'human nature'" (Machery 2016a, 210); Samuels cares for mechanistic explanations in cognitive science; Griffiths and Stotz care for developmental biologist's business. Ramsey (forthcoming) is also concerned with "how traits come about" developmentally and presents his account as tracking the associations that can then be explained mechanistically in Samuels's sense. In order to have a more general account that includes a larger variety of contexts, one needs to take a step back to

Hull's basic point: human nature refers to important descriptive traits of our life form. The issue that needs to be answered is then what *important* means. The answer will (again) differ with the epistemic contexts. Carrying a cell phone is quite important in some contexts, as when a scientist aims to explain how humans communicate and function socially. Yet it is quite unimportant if our evolution is at issue.

Thus, my claim is that importance is the second qualifier. Whether it means stability (as I suggest for the context of evolution, development, and heredity) or something else depends on context. Which generalizations are regarded as important for the respective life form might thus well vary across cultures and epistemic contexts without this causing a problem for a shared concept of human nature, except that the content varies, as already in the vernacular concept, but this time without taking away objectivity.

The objectivity stems from the objectively established list of typical properties from which one picks—relative to scientific interest—one or a set of properties. Since there will always be a plentitude of properties that are typical among humans, talk about human nature will involve a choice of those properties deemed to be relevant in a certain context. As mentioned elsewhere (Kronfeldner, forthcoming), an interesting underdetermination results: different scholars or scientists can prioritize certain properties without science giving any foundation in objective facts for the priority chosen. Some researchers will highlight rationality, others morality, others our violence, still others the opposable thumb, or some other trait. As in all cases of underdetermination, not only disciplinary focus but also social values will govern the choice of what is most important for being human, with often no scientific way of finding agreement on the choice. Yet that does not entail that the list of typical properties itself, from which a subset is chosen, is not objective.

My answer to the problem of whether typicality is sufficient is thus as pluralist as my answer to how to apply typicality in face of the fact that all typical traits involve abstraction from details. It depends on whom one asks, that is, on the epistemic context.

In the following, I will offer three different kinds of examples of choices from the list of typical properties: there are, for instance, the philosophically minded ones that focus on uniqueness and importance in our self-interpretation (how we want to be); there are explanatorily minded scientists who focus on traits of our life form that explain other traits of our life form, traits that are important in the sense of explanatorily central; finally, there are those who care for a certain kind of stability as the decisive criterion, in line with the assumption that the concept of human nature assumes some fixity of the traits.

I will then show in chapter 7 that within this third context, which is predominantly but not only an evolutionary context, the best characterization of how to narrow down the list of typical traits (to exclude properties such as carrying cell phones from human nature) is still by use of the traditional nature-culture divide because it allows distinguishing among different kinds of stability. The way I delineate nature from culture is different from other proposals, for example, Machery's, who is, among those involved in the debate in philosophy of science, nonetheless my closest ally in that respect.

Uniqueness as Epistemic Interest

Most philosophers (but also others) will take *importance* to refer to those properties that are typical and unique for humans (i.e., species specific). Some will do so (as Ramsey 2013) because they realize that their account is otherwise too inclusive; others (e.g., paleoanthropologists) will do so because they are interested in a classificatory question (which individual belongs to the species and which does not). Others (e.g., philosophers) will search for unique traits because they want to describe in which sense humans are special, that we are not just another animal. They are interested in our self-interpretation, which connects to how we want to be. Since that self-interpretation will itself vary with the epistemic, social, and political context, the choice of traits varies with context and might even conflict, depending on which evidence is highlighted. Milam (forthcoming), for instance, shows how the experience of the atrocities of World War II influenced a stark conflict in our self-interpretation: a conflict between a utopian picture of human evolution, which attributed our uniqueness and evolution to creativity and cooperation (as, for instance, in Dobzhansky 1956), and a dystopian "killer ape model" that takes violence and killing of species members as the self-evident and unique trait of the human life form (as, for instance, in Lorenz 1960). Milam also shows that the different self-interpretations in the second half of the twentieth century not only depended on different reactions to the political background but also on the different kinds of evidence used: evidence from fossils, current hunter-gatherer cultures, or animal behavior. Depending on the choice of evidence, something else about us becomes visible. The conflict between the utopian and dystopian image might thus have been epistemic only: explainable by different epistemic focus. Consequently, a unique trait such as upright gait or a specific teeth form will be important for paleoanthropologists dealing with fossils; an ethologist, primatologist, or cognitive scientist will focus on her or his own evidence (e.g., certain behaviors in current animals) and will rather focus on territoriality or similar behavioral traits.

Philosophers, finally, will focus on traits such as rationality, morality, and the like. And even they will disagree about which trait is most important. Crane (2013) stresses the "pursuit of knowledge," while Olafson (1995, 12) stresses a Heideggerian "familiarity with being as such." Such a disagreement among philosophers does not stem from one of the traits being more or less typical, but from the fact that they look at human nature from within a certain philosophical paradigm. Disagreement will also result from a difference in the kind of importance at issue: metaphysical importance, political importance, or cognitive importance, for example. For the last, psychologist will be keen to join in, but will add some explanatory interest.

Explanatorily Central, Since Structural

Having a language, for instance, is usually considered part of human nature. It is of central cognitive importance since it is not just typical and (in its narrow sense) unique to humans (as a population), but also explanatorily relevant for other traits that rely on it, be these traits typical (e.g., morality) or varying (e.g., speaking Japanese). Since there are explanatory relations between the properties of a cluster, Roughley (2011) limits the descriptive human nature to a set of properties that are explanatory for other traits and central in that sense. The kind of explanation he points to, however, is not causal. According to Roughley (2011), traditional claims that such traits constitute human nature are, once divested of any definitional aspirations, best understood as claims that these traits are "structural properties of the characteristic human life form." Consequently, they are structurally explanatory—structural in the sense that it is about the relations between properties of the human life form. The more central aspects of the life form are necessary for the less central ones to be possible. Intentionality, for instance, is structurally central: without it, humans would not have the complex language and culture that is characteristic for them.

As Kronfeldner, Roughley, and Toepfer (2014) stressed, one of the best-known recent empirical accounts of the way humans are is provided by Tomasello, a psychologist. His preferred level of analysis is that of individuals, humans, as he mostly calls his subject of investigation (e.g., in Tomasello 2008). He does not use the term *human nature* and only very rarely calls humans a species. Nonetheless, the empirical research in Tomasello's case does two things: it picks out typical traits of the human life form and establishes explanatory relations between them. The comparative psychological tests that Tomasello and his colleagues conducted with humans and other primates led to the claim that one trait is distinctive and explanatory for human social interaction: shared intentionality, the ability to participate with conspecifics in collaborative activities with shared intentions, a mutuality that is not only typical of but also specific to humans. Connected with and explained by shared intentionality is a pattern of capacities, psychological dispositions, and motivational states that are equally typical and specific for humans, such as our language capacity and the cumulative culture that is so obviously unparalleled.[12]

The main problem I have with this is that it is unclear how an abstraction, such as the abstract property of "having a language," can explain what it is abstracted from, namely, concrete instantiations of languages—somebody speaking Japanese, somebody speaking English, and so on—even if one assumes a concept of structural explanation. In addition, biologists will regard something else as central. They are likely to count our "being mammalian" as of utmost importance for our nature, while philosophers as well as scholars from the humanities and social scientists presumably could not care less about that part of our human nature. Explanatory centrality depends on the explanandum chosen.

Finally, some traits will be more central, some traits less central and therefore also more or less part of human nature. Yet this is a problem other accounts in the literature face too, as I shall argue below.

Fixity, or Rather Stability?
Within the philosophy of science literature on human nature, a certain stability, if not fixity, has often been added as a further ingredient to what a descriptive concept of human nature needs. The post-essentialist account

that I defend builds on that aspect but explicates stability with reference to genealogy and adds, once more, a pluralist twist.

I have said above that human nature in the descriptive sense refers to a set of typical and stable traits of the human life form, a cluster of properties that extends in space and time; it is a set of properties that are typical for humans at a certain point in time and for a considerable amount of time. In other words:

> Descriptive human nature = properties that are instantiated by a statistically significant number of humans that reliably reoccur over a significant time span.

The important typical traits of a life form are those that are stable. The stability accounts for the stasis of the species and thus for the fixity that I mentioned in chapter 1 as one of the building blocks for concepts of human nature. Stability, though, in contrast to fixity, allows for change. Evolution is about change, but it is also about stability within change. Stability within change is what one needs to look at in order to understand the descriptive concept of human nature in a manner that allows reconstructing what remains from the old fixity.

I do not know of any author who would challenge the claim that there are such stable generalizations about humans that can be used to describe the human species. The respective stable generalizations (e.g., that humans have language) are often conceived as resting on what Boyd (1991, 1999) has called homeostatic mechanisms, which I discussed in chapter 5. The phenetic clusters, the set of typical (not necessarily specific) phenotypic properties, are accordingly called homeostatic property clusters. Differences exist on how to spell out exactly what Boyd calls homeostasis, Griffiths (1999) phylogenetic inertia, and I stability, but I do not think these differences matter. What does matter and is contested is how the stability relates to further qualifiers such as innateness and evolvedness. This is the subject of the next chapter.

7 The Stability of Human Nature

In this chapter, I discuss how stability can replace the fixity that is often associated with human nature in the descriptive sense. This will bring to the fore a post-essentialist solution for the developmentalist challenge of how to make sense of the divide between nature and culture, given the so-called interactionist consensus described in detail in chapter 4.

The two main negative claims are that innateness is not adequate to reconstruct the fixity implied in the concept of human nature since it refers only to the developmental level; evolvedness fails because it fails to exclude cases that are justifiably regarded as not being part of human nature. The positive claim is that the best replacement for fixity is the stability that the biological channel of inheritance conveys. This solution uses, the nature-culture divide in the sense of the channelism introduced in chapter 5, abstraction, and the difference between explaining traits and explaining trait differences. The last two also allow reconstructing what it could possibly still mean to say that something is exclusively "due to nature" or "due to culture."

In section 7.1, I discuss and finally dismiss innateness and evolvedness as options to narrow down the descriptive concept of human nature. I present channelism in section 7.2 as the foundation for a solution that refers to biological inheritance. In addition, the language of traits being "due to nature" or "due to culture" is defended as referring to something real. Given the solutions suggested, a narrow enough concept of a descriptive nature is presented in section 7.3 as resulting from the claims made in chapters 6 and 7. The result is a midpoint between throwing out the baby with the bathwater (no trait is part of human nature) and drowning the baby (everything is part of human nature).

7.1 Innate or Evolved?

Innate

Griffiths (2002) implicitly assumes that the concept of human nature entails the concept of innateness. Since innateness refers to developmental fixity, human nature does too. The concept of innateness is historically related to the concept of human nature but far from equivalent to it. Innateness relates to the intuition that what is part of human nature is also "given" to an individual and "hard to change," an aspect that John Stuart Mill (1874) lucidly analyzed and that is hotly debated in discussions on enhancement.

The problem is that traits (e.g., a certain genetic disease) can be innate without being part of the shared human nature. If at all, such traits are part of what one might want to call an individual's nature, irrespective of a shared species nature. Yet for a long time, people would have agreed that traits that are part of human nature are innate. But even that is now contested since the concept of innateness is almost as contested as the concept of human nature, given the interactionist consensus. In response, an industry of successor notions for innateness has developed. Mameli and Bateson (2006, 177) list "twenty-six candidates for scientific successor to the folk concept of innateness," one even reducing innateness to "species-typical," which evidently will not help solving the problem at issue in this study.[1]

A simple specification of innateness is as "arising from genetic factors," as Fukuyama (2002), for instance, has it.[2] He consequently specifies human nature as follows:

> Human nature is the sum of the behavior and characteristics that are typical of the human species, arising from genetic rather than environmental factors. (Fukuyama 2002, 130)

This assumes that there are traits that exclusively arise from genes, which conflicts with the interactionist consensus. The latter entails that genes always interact with other developmental resources so that traits never arise from genes alone, and it also (by now) entails that there are epigenetic factors involved in biological inheritance.

It follows that if "exclusively arising from genetic factors" were treated as the second criterion, in addition to typicality, then human nature in the descriptive sense would vanish (rather than becoming too inclusive).

Human nature would simply become not just thin but an empty class since everything arises from genetic factors interacting with epigenetic, behavioral, cultural, and environmental factors, and therefore nothing could be part of human nature. Equating the fixity of human nature with "arising from genetic factors" is flying in the face of the interactionist consensus.

Innateness can, however, also be reconstructed and interpreted as meaning phenomena like adaptive entrenchment or developmental canalization. Ramsey's (2013) robustness is a variant of such a specification. These much-debated replacement candidates for innateness do not conflict with the interactionist consensus. Yet there are two problems with regarding human nature as consisting of traits that are typical and innate, even if innateness is specified as entrenchment or canalization. First, innateness in an interactionist sense is gradable. Second, innateness and evolutionary stability are not the same.

First, innateness, in the sense of canalization or entrenchment, does not allow for a clear boundary between innate and not innate. It can even be counted as being the point of the successor notions to go beyond that dividing line in order to accommodate the interactionist consensus. Traits are more or less entrenched or canalized, for example. But are traits more or less part of human nature? It might well be that one has to bite that bullet in order to arrive at a post-essentialist concept of human nature that, on the one hand, meets intuitions about certain examples (e.g., language capacity as belonging to human nature and cell-phone-carrying-habit as not, despite both being typical) and that, on the other hand, is compatible with contemporary biological knowledge, in this case mainly the developmentalist challenge.

Second, and more important from my point of view, traits can fail to be developmentally fixed (i.e., be highly dependent on environmental input, be plastic in any sense one might wish), and still the trait might reliably reoccur over evolutionary time in the species. The human language ability, a property that gets realized in different phenotypes differently, is a case in point. This relationship between high evolutionary stability and high developmental plasticity has been dealt with intensely, most prominently by West-Eberhard (2003). This also shows that innateness refers to an individual and developmental context, whereas human nature refers to a populational and evolutionary context.[3] That is the main reason that innateness will not provide a solution for specifying a successor notion for the essentialist fixity of human nature.

Evolved?

Machery (2008, forthcoming) is therefore on the right track when he criticizes the concept of innateness.[4] Consequently, he ignores it in his suggestions for a successor notion of fixity: he instead refers to human nature traits as traits that are typical and evolved. I take him to cover with this the same aspect of the concept of human nature as I intend here: that properties that are part of human nature are somehow more stable than others—more stable not just (and not necessarily) developmentally (i.e., in individuals) but evolutionarily (i.e., across individuals).

In the framework I am using, it holds: if human nature traits are biologically evolved, then that can explain the high stability in time and space of traits that belong to human nature. Yet, and that is the complexity ignored by Machery, the reverse is not true: one cannot infer from a certain stability that a trait has biologically evolved, since it might have evolved culturally. I therefore claim that what counts for descriptive nature is that a trait is stable and not that the trait is evolved, since it could be made stable by a different process, say, creation. Conceptually, a creationist can believe in the same concept of human nature as an evolutionist; the typically and stably reoccurring traits of the human life form are part of human nature, a creationist would say—not because of evolution but because the creator intended the world that way. The concept of human nature is the same; it is just connected to a different kind of explanation. I first analyze Machery's proposal in more detail, which will lead back to the channelism from chapter 5.

Machery's Proposal

According to Machery's (2008) account of human nature, something that is typical and "due to nature" and therefore "part of human nature" (rather than culture) is something that is the result of evolution, that is, something that can be explained with an ultimate, evolutionary explanation (rather than being subject to a proximate, developmental one only).

The contrast between ultimate (roughly, evolutionary) and proximate (roughly, developmental) explanations, a distinction introduced by Mayr (1961), has its difficulties. A dominant critique against the distinction between proximate and ultimate explanation is that often (not necessarily always) an ultimate explanation will include a proximate explanation for the explanation of why the selective situation, on which the ultimate explanation draws, was the way it was. In such cases, the *how* (proximate explanation) is entangled in the *why* (ultimate explanation)—a developmental explanation entangled in an evolutionary explanation. Developmental systems theory has defended this direction of entanglement ever since.[5] Laland et al. (2013) also stress this direction of entanglement forcefully with respect to insights from evolutionary developmental biology, niche construction, human cooperation, and cultural evolution. Proximate mechanisms can "modify the selection acting on individuals and thus must feature in evolutionary explanations" (Laland et al. 2013, 723). In many cases, they claim, causation is reciprocal in the sense that the answer to how a system works often influences why the system gets selected, which influences further rounds of selection, ad finitum. Development, culture, and niche construction are thus evolutionarily, not just developmentally, relevant.

Given these critiques, Machery's account of human nature seems to rely on a flawed proximate-ultimate distinction. Although I agree that Machery's account fails, I do not think that it fails because of how he uses the distinction. First, even if the two kinds of explanation are entangled in the sense discussed in chapter 4 and above, this does not mean that the analytic distinction between ultimate and proximate explanation does not make sense. Asking how and asking why are still two different ways of asking for an explanation, each ignoring different aspects of a phenomenon. Second, so far, only one direction of entanglement has been shown. Is an evolutionary explanation always (or at least often) part and parcel of a developmental explanation? It has not been shown that the why is equally often entangled in the how and this is why Machery (2008) is immune against the critiques I have noted. He stressed that there can be traits that do not, or at least not yet, have an evolutionary explanation despite the fact that everything somehow has a causal evolutionary history (i.e., would not be here without evolution). Even the belief that water is wet is, according to Machery, a result of evolution, since without evolution, there would not be a mind that has that belief. Yet he maintains that such a belief itself does not have an evolutionary explanation. What that means is simply that it does not *yet* have an evolutionary explanation since the belief itself does not yet have an evolutionary history. So he is defending an independence of some proximate explanation from ultimate explanation rather than the independence of ultimate explanation from proximate explanation. If there are traits that have a proximate explanation without yet having an ultimate explanation, there is indeed an important explanatory asymmetry involved that justifies using the proximate–ultimate distinction.

Although Machery can react in the just-specified way, he needs some way to explicate that there are traits that do not yet have an evolutionary explanation, despite somehow being the result of evolution. In a reply to critics (Machery forthcoming), he provided such an account. He uses a contrastive account of explanation (with reference to Van Fraassen 1980 and Lipton 1990) to divvy up properties explained by culture and those explained by nature. According to that account, it is possible and plausible to regard some traits as not having an evolutionary explanation, despite the dependence of every trait on there having been an evolution of the system exhibiting the trait. "For instance," he writes,

> the fact that Americans can hear and enjoy music fails to explain why they know the "Star-Spangled Banner," because it fails to explain why they know the "Star-Spangled Banner" *rather than some other song*. For the same reason, adding the fact that Americans can hear and enjoy music to the fact that the "Star-Spangled Banner" is the American national anthem does not improve the quality of the explanation of the fact that Americans know the "Star-Spangled Banner." (Machery forthcoming, 31–32, emphasis added)

I think this is on the right track: the explanation of knowledge of the Star-Spangled Banner needs to be specific, that is, proportional to what exactly is in need of explanation. A general explanation of why Americans are able to hear music, whether their national anthem or any other song, is not adequate since there is a more specific explanation. Yet such a defense needs a more secure pragmatic-pluralist foundation, which I will provide. As a step toward it, I first provide a solution to the preliminary question, having to do more with description than explanation, about how to narrow down the list of typical traits in order to arrive at a sensible concept of a descriptive nature.

For that, I think the best interpretation (or revision) of Machery's account is one that does not use any reference to proximate and ultimate explanation. Here it is: Machery's nomological concept of human nature refers to traits that are typical and conserved by evolution. To be more precise, human nature consists of those traits that are typical because they were conserved by evolution. It is important to note that it is irrelevant for Machery (2008, 323–324) whether the trait is conserved by natural selection, sexual selection, drift, or what have you. To include only traits that have been selected for due to natural selection rather than one of the other selection processes would be too narrow. It would exclude traits that are part of human nature without being adaptations. A good candidate is the fact that our bodies are symmetric: part of our human nature but unlikely to be an adaptation.

Although I think this is correct, there are two problems with Machery's account: the first problem is similar to the vagueness problem that innateness faces, and the second problem relates to the goal that the list of traits belonging to human nature cannot be sufficiently narrowed down that way. Machery's account (even in that revised version) fails to exclude traits like our cell-phone-carrying habit and other, potentially much older, cultural traits such as knowledge of how to make fire or the habit of burying the dead, which are, at least in some scientific contexts, not taken to be part of human nature since humans learn them. They are part of human culture. "Typicality" plus "conserved by evolution" seem to be still insufficient to narrow down the list of traits so that folk intuitions about "carrying a cell phone" and "burying the dead" as not being part of human nature are reconstructed adequately. I first address the vagueness problem and then the narrowing problem, which is the harder one.

The Vagueness Problem Regarding "Evolved"

How old and conserved by evolution must something be in order to count as "due to" evolution and therefore "part of human nature"? Hard to say. Think about a mutation leading to a new trait variant that had such fitness-enhancing effects for humans that it spread and became typical within a very short time (give or take a hundred years). Is the resulting trait then old and thus evolutionarily conserved enough to count as part of human nature? Or is it still a "sport," and regarded to be so heavily scaffolded by culture that it would not be there without the latter?[6]

Consider, in addition, a very old trait, clearly evolved but rather untypical: lactose intolerance, which was quite typical for humans—the original condition, so to speak—but it is no longer typical. It has been replaced in some cultures, at least in the West, as a result of the cultural evolution of agriculture, with lactose tolerance. The replacement is so prominent that, as described in chapter 4, the evolution of lactose tolerance counts as the paradigmatic example for a coevolution of nature and culture. Now, what is part of human nature: lactose intolerance or lactose tolerance? Lactose intolerance is the older, original, and evolved condition, and thus part of our nature, right? Yet it is far from typical now, and many consider it a disease. So now it is lactose tolerance that is part of human nature, right? Well, some human groups are tolerant regarding lactose, and others are not. Lactose tolerance is not yet widely distributed (prevalent in all human groups), while lactose intolerance is no longer widely distributed either. If "evolved" (and thus "old") carries significant weight, the nomological account is forced to say that lactose intolerance is what is part of human nature. It was once typical and is still around, never mind the feelings of those who get diagnosed (or self-diagnose) a by now quite fashionable "disease" called "lactose intolerance." In a nutshell, the vagueness issue about stability (typicality in time: how old is enough) is just the mirror image of another vagueness issue: How typical has something to be to belong to human nature? Is 70 percent enough? Again, hard to say.

There are no all-purpose answers to vagueness issues. Delineating human nature as a set of typical and stable properties will be a dirty, fuzzy business for both criteria, typicality and stability. But as long as one is not in the business of delineating necessary and sufficient conditions, this does not in and of itself count against a proposal. As long as the criteria can fulfill the epistemic roles they are meant to fulfill, vagueness is not a problem. For the descriptive role, which is the role at issue in this chapter, the vagueness problem is a problem only in principle. In practice, agreement on how much is enough is the norm rather than the exception. Yet this is an estimate. Only empirical case studies on respective scientific fields would allow saying more on the question of how and how often scientists exactly solve these vagueness issues. Thus, my reply to the vagueness issue is as with the polymorphism issue: the details of the epistemic practice at issue, in science or elsewhere, are key for deciding what counts as part of human nature, but typicality and stability are among the best criteria to make the decision. Nevertheless, one important problem remains.

The Nature–Culture Divide and Kinds of Stability

There are typical traits that are quite old and conserved by evolution, that reliably reoccur across generations but are nonetheless standardly regarded as not being part of human nature since they are part of culture. I already mentioned the habit of burying the dead as a case in point and will use it

The Stability of Human Nature   155

to clarify the most important problem Machery faces. The example shows that typicality plus being conserved by evolution are not yet sufficient to narrow down the list of traits. We need an extra criterion, one that uses the nature–culture divide in order to specify different kinds of stability.

Machery (2008) implicitly seems to equate "evolved" with "biological inherited" when he writes:

> Saying that a given property, say a behaviour, such as biparental investment, or a psychological trait, such as outgroup bias, belongs to human nature ... is also to say that some kinds of explanation for the occurrence of this trait among humans are inappropriate. Particularly, this is to reject any explanation to the effect that its occurrence is exclusively due to enculturation or to social learning. (Machery 2008, 326)[7]

The problem is that with this, one can get rid of the cell-phone-carrying habit as part of human nature, but one would still have to include the habit of burying the dead as part of human nature. The problem is that social learning is interpreted as being short-lived only (and therefore as figuring only in developmental rather than evolutionary explanation). But culture is far from so short-lived, as I demonstrated in chapter 4. It can be quite long lasting, evolutionary itself, and stability inducing. In the terminology of proximate and ultimate explanation, it is an ultimate force too, as Laland et al. (2013) and Mesoudi et al. (2013) stress.

Evolved versus not evolved (in whatever interpretation) will simply not match nature versus "exclusively due to social learning." Given cultural evolution, the qualifier "evolved" is simply too broad to help solve the remaining problem of narrowing the list of traits (given that typicality is already established).

I regard it as a safe assumption that there simply are traits that are old, evolved and whose presence is explained by developmental resources that are transmitted culturally rather than biologically, for example, the habit of burying the dead, which is rarely considered as part of human nature in the narrow sense. It is standardly rather considered as what Dobzhansky (1956, 118–122) called "common denominators of culture." Like Machery, I assume that culturally transmitted practices or knowledge should not be considered as being part of human nature because they involve cultural inheritance. Yet I do not equate evolved with biologically inherited since the habit of burying the dead, a rather typical human trait (not strictly species specific, to the best of my knowledge), is clearly evolved but standardly regarded as not being part of human nature because it is explained differently. The trait would not be typical without social learning, and still it holds that to explain the trait (with its contrast: to not bury the dead) evolutionarily as well as developmentally, one will always have to involve an evolutionary process of variation, plus cultural inheritance and selection, leading to its retention over time. But one would not therefore regard the trait as part of human nature.

The latter is important since Machery himself does recognize that certain traits can be typical and around for a long time without being part of human nature. For him, being around for a long time is not the same as "evolved" (conserved by evolution):

> Saying that a trait has an evolutionary history is something stronger than the fact that it has perdured across generations. Humans have probably believed that water is wet for a very long time, although this belief has no evolutionary history. For this trait is not a modification of a distinct more ancient trait. (Machery 2008, 327)

This might well be true for the belief that water is wet, but it is very likely wrong for lots of culturally transmitted traits, for example, the habit of burying the dead. I take it as safe to assume that it involved cultural variation, cultural inheritance, and cultural selection. Machery wrongly denies the possibility of cultural evolution over long evolutionary time, as Lewens (2012) already criticized. As described in chapter 4, the possibility of a separate cultural evolution (which can also lead to a coevolution of nature and culture) is broadly acknowledged by now.

From Machery to Channelism

Building on what I have said already on the evolutionary interaction of nature and culture in chapter 4, I will defend distinguishing between the two in the next section and thus establish a narrow enough concept of a descriptive nature. That there is a separate cultural evolution allows delineating different kinds of stability. One of these kinds of stability is exhibited by traits that are part of human nature, and another kind is exhibited by traits that are part of human culture.

The stability criterion that I use is similar to Machery's evolvedness criterion, but since it is directly based on the qualifier "biologically inherited" rather than "evolved," it prevents the two problems Machery faces: it is irrelevant how old a trait is; as long as the trait is typical and explained by developmental resources that are biologically inherited, it is part of human nature; evolved cultural traits are clearly excluded. But I side with

Machery with respect to Lewens's (2012) claim that "it is not the role of a definition of human nature to exclude social learning or enculturation as the explanation (or, for Machery, the sole explanation) for the widespread development of some important feature of our species" (466). I think all the language of X being "due to nature" and the history of the nature–nurture (and nature-culture) divide speaks against Lewens's dictum, despite the developmentalist challenge. In reply to Lewens (2012), Machery (2012) opts (as I do here) for distinguishing between "organic evolutionary processes" and "cultural evolutionary processes." Still, Machery has not offered a detailed solution on how one can make that distinction precise, given cases like the habit of burying the dead. I will provide one in terms of the channelism introduced in chapter 5 and in terms of the in-principle legitimacy of abstract traits (taken as explanandum phenomena), as introduced in chapter 6. Thus, although my account differs from Machery's, I take him as an ally in defending different channels of transmission, against Lewens and the holists from chapter 4 who argue against it.

7.2 Channelism, Stability, and the Nature–Culture Divide Revived

The Stability of the Biological Channel

Developmental resources have to endure in order to reliably cause over time the properties typical of a species. Their endurance causes coherence and stasis of a species and, via this, the stability of the respective property cluster. There are, as mentioned in chapter 5, different ways to endure over time. If things do not simply persist (such as mountains separating two populations), they have to be handed down (i.e., transmitted) to the next generation by biological transmission (as genetic and epigenetic factors) or cultural transmission. The idea of cultural evolution rests on three assumptions: that there is (1) some autonomy of the variation inherited through the biological channel from the one inherited through the cultural channel, (2) near-decomposability of the channels, and (3) a distinct temporal order inherent in the channels. For this chapter, the important aspect is the third one. I justified it in detail in chapter 5 and briefly summarize the result here.

The channel of transmission makes a dynamic difference to the evolutionary stability of the developmental resources. The biological channel rests on genealogy (i.e., the vertical relation between parents and their descendants) and the mechanisms of biological heredity. The mechanisms of biological heredity make transmission mostly content independent and thus very stable. Culture, by contrast, is content dependent and its stability varies a lot.

Applied to the problem of finding a narrow enough concept of human nature, this means that only those traits that are typical and are accounted for by developmental resources that travel the channel of biological inheritance count as part of human nature in the descriptive sense.

But are there such traits—traits that are due to biological inheritance or due to cultural inheritance and therefore part or not part of human nature? Isn't that dangerously close to Fukuyama's traits that "arise from genes alone"? I first illustrate how traits can be due to cultural inheritance and then how they can be due to biological inheritance despite the interactionist consensus.

How Nature Averages Out

Traits that are due to cultural inheritance are traits to which biological differences (differences in biologically inherited resources) are not making a difference, even though, as described in chapter 4, biologically inherited factors always need to be there to produce any phenotypic trait whatsoever. The key difference between my account and Fukuyama's (and similar other accounts) is that my account uses the distinction between difference making and production of a trait, as introduced in chapter 4. (This is also the extra bit that Machery needs in order to get his contrastive explanation answer really off the ground.)

There are indeed such situations where nature does not make a difference for an effect—situations, where, at the populational level, nature averages out. Alfred L. Kroeber (1917) long ago pointed to such situations, first with respect to human nature as not making a difference for cultural change, given that the latter is autonomous in the sense specified in chapter 4. For instance, as described in that chapter, individual biological differences seem not to make a difference to the averaged difference between body height these days and averaged body height five hundred years ago. By choosing the explanandum accordingly (averaged differences in body height among all humans as trait), these biological differences can be regarded as randomized and neutral, even if—as a matter of fact—individuals differ with respect to the biologically inherited developmental resources

The Stability of Human Nature 159

(labeled as alleles A and B in figure 4.2) that are involved in the production of individual body height. The historical difference in averaged body height is thus a purely cultural effect, or a cultural trait.

Also synchronically, individual differences in biologically inherited developmental resources can sometimes be ignored in order to account for a difference with respect to a trait at issue. Even the developmental holists, among the hardest critics of the nature-nurture divide, allow for that kind of irrelevance of biologically inherited resources. Griffiths and Gray, for instance, write:

> The phenomena of habitat imprinting demonstrates very nicely how the association of an organism with an environmental feature could have an evolutionary explanation *without the genes having an interesting role in the production of that trait*. ... The habitat is something they [populations of the European mistle thrushes] have acquired through evolution, as much as any other element of phenotype. Yet the genetic variation between the two populations can be presumed to be random with respect to which habitat they have imprinted on. (Griffiths and Gray 1994, 288, emphasis added)

Although they use the language of producing rather than difference making, which is quite misleading, they basically admit that in the case at issue genetic variation is not making a difference for habitat imprinting.[8] The genes do not have an interesting role in the "production" of the traits at issue since they do not make a difference, that is, they do not account for the difference in habitat imprinting of the two populations at issue. Such a difference can well persist over time, or even be inherited and selected. In these cases, genetically inherited developmental resources do not make a difference. The differences regarding the trait can then be regarded as due to nongenetic developmental resources alone. If these nongenetic developmental resources are socially transmitted, then the respective trait, difference in habitat imprinting, is due to that kind of transmission alone.

One might want to object that, as mentioned in chapter 4, a difference regarding a trait is not the same as the trait. True, but any difference regarding a specific trait can be revamped as another trait. Here are two examples. One is the example from chapter 4 of the different differences the biologist and the anthropologist focused on. Some (e.g., Fukuyama) were interested in why men and women have (on average) different body heights rather than the same; the cultural anthropologist or historian was interested in why humans (all together, on average) show a different body height today compared to five hundred years ago. One only needs to give these different differences (explananda of the different scientists) fancy-sounding scientific labels, for example, "sex-related height effect" for the first (the difference the biologist is interested in) or "effective body height" for the latter (the difference the anthropologist is interested in), so that the new explananda sound more like traits. Each is caused respectively by nature alone or nurture alone, whereas the original trait, body height pure and simple, is still caused by nature *and* nurture.

Language is another example: people differ regarding the concrete languages they speak. Some speak English, some Japanese, some this, some that. The differences in their language are exclusively due to differences in developmental resources traveling by cultural inheritance. Which language one speaks is thus due to culture alone. This holds even though biologically inherited developmental resources are necessary to produce organisms that have a language. None of the individuals would speak either of the concrete languages without also having had thousands of developmental resources available that travel through the channel of biological inheritance. The difference in their language (language now taken as something abstract) is an abstraction from the comparison of the concrete languages they speak (similar to the "effective body height" as an abstraction from the differences in actual body height between the fifteenth and the twentieth centuries). Only in that sense can the trait "speaking English rather than Japanese" (or any other concrete trait with a contrast) be said to be due to culture alone. As in the habitat imprinting that Griffiths and Gray mentioned and my height example, biologically inherited resources are not making a difference to the differences—in habitat imprinting, in height, in language—of the one population compared with the other. Since these trait differences can be revamped as another trait, there are traits that are due to culture alone.

How Culture Averages Out

In a similar manner, a trait can be regarded as due to nature alone. All one needs is some kind of averaging out. Every person speaks a specific language: some speak Japanese, some Chinese, and so on. By using the kind of disjunctive solution discussed in chapter 6 with respect to polymorphisms, one can say that, in abstract terms, they all nonetheless share one trait: they have a language. In concrete terms, by contrast, nobody speaks *a* language,

The Stability of Human Nature

since everybody speaks either, say, Japanese or Chinese. To get traits that are due to nature alone, one simply uses the method of ignoring differences until one reaches a situation and a concept of an abstract trait with respect to which culture no longer makes a difference.

One can even play the game of highlighting or ignoring differences to have effects that are due to one kind of causal factor alone. Imagine a gene known to be relevant for language development (e.g., the FOXP2-gene, the putative "language gene," coding for the Forkhead box protein P2).[9] Imagine that it influences whether an individual can develop a full-blown spoken language or suffers from some severe limitations. Add an intuitive grasp of the cultural influences necessary to develop the ability to speak a specific language, for example, Japanese. Now imagine that one can compare some of these cultural influences (developmental resources handed down via social learning) so that one can convert them to one currency and line them up in a quasi-quantitative comparison (similar to the nutrition from fifteenth to the twenty-first centuries in the height example). Imagine then, for illustration, a color wheel, and the differences in concrete languages matched to the differences in color on the wheel. If one speaks Japanese, one gets represented as a green dot; if one speaks English, as a blue dot; if one speaks Ubuntu, as a red dot. A dot can be in the inner center or the outer plane of the wheel. A full-blown language capacity is in the full-blown outer range. Although the production of every individual's specific language ability is due to nature and nurture, it is still possible to say that the FOXP2-gene makes a difference to the difference between having a full-blown spoken language or not and the environmental factor makes a difference to differences in specific spoken languages, that is, a difference to which specific language an individual speaks (regardless of whether it has the FOXP2-caused impairment).

Certainly this finding of two kinds of independences regarding differences in a trait holds only for the measured situations. After all, the problem of extrapolation (described in chapter 4) also holds for this case and all other cases discussed in this section. The independencies only hold ceteris paribus. For instance, FOXP2 is believed to be causally important not directly for language itself but for developing motoric abilities necessary to speak a language. There are then very likely also other factors that are equally necessary for developing the respective motoric abilities. For instance, the development of the muscles in the throat, evidently necessary for spoken language, seems to depend on a peculiar environmental influence, namely that babies chew hard things at a certain time in their development. Thus, there will be patterns of dependence that complicate things and will make it hard to defend the claim that only nature (biologically inherited factors) makes a difference to whether the individual develops a full-blown ability for spoken language. What is part of human nature would then be something even more abstract than "speaking a language." Thus, given new evidence, one might be forced to move even further away from the concrete abilities. Imagine new evidence about environmental factors influencing the development of the grammatical competences in children. What is part of human nature would move further toward the abstract, toward something like "acquiring some kind of linguistic competence."[10]

The more abstraction is involved, the stranger the trait sounds and the less justified it is to include such a trait as part of human nature. The language case would then approach the hypothetical case from chapter 6 of "having something like some fingers" as part of human nature.[11] It could formally count as part of human nature. In certain contexts that might still do some epistemic work, producing some interesting knowledge, but in other contexts, the abstraction might be so forced that it becomes a practically useless exercise to establish the abstract trait. In certain contexts, it might be better to look instead at the culture-dependent diversity.

Yet this does not diminish my claim that by taking abstraction into account, one can make sense out of people claiming that "having a language" is due to nature alone and "speaking Japanese" is due to culture alone. In the imagined language example, it would mean that the gene(s) make a difference to whether it is spoken full or rudimentary, but do not make a difference to which language is spoken.

To sum up, to make sense of the language of a trait being "due to nature" (or nurture, or culture) is, first, to make the contrast (i.e., difference) implied in the explanandum explicit, as Machery already suggested; second, to see the difference between a trait explanation and a difference explanation; third, to accept abstraction, that is, the reconstitution of phenomena via abstraction as legitimate in principle. (Machery's, forthcoming, suggestion to use a contrastive account of explanation uses only the first point.) Certainly, whether, for a given trait, there are such differences to which either nature or culture is not making a difference (in concrete or via abstraction), that is, which patterns of dependence there are is in each case an empirical matter, and certainly always subject to the limits outlined in chapter 4, in particular, with respect to extrapolation.

Before I can come back to the results regarding the question of whether typicality and stability taken together give us a narrow enough concept of human nature, I draw a further conclusion about the legitimacy of using disjunctions.

The Legitimacy of Using Disjunctions

The abstract property "having a language" is—given current evidence—one of the least contested examples for a trait being part of human nature. But look at how it has been established: via a disjunction! First, it was observed that humans in actual, concrete terms either speak this language or that language or another language (as one observes that most humans have either the female sexual morph or the male sexual morph typical for humans). Then one takes into account that some do not speak their language but rather sign it or use still other means (as one observes that there are some who are intersex). From that diversity, a common trait is abstracted—having a language.

This brings us back to the discussion of polymorphisms in chapter 6. Machery (2008) is not willing to include disjunctive properties such as polymorphism (e.g., sex-specific properties), while he is probably not hesitating to include the disjunctive property "having a language." Is that a problem? I do not think so since there is no general and a priori answer where abstraction should stop (i.e., where the disjunctive regress should stop). In addition, the regress goes both ways: on the one hand, there are traits that show a high diversity and are thus standardly not accepted as part of human nature but could be accepted via abstraction; and on the other hand, there are traits (such as having a language) that are (radical behaviorist notwithstanding) standardly taken as uncontestably being part of human nature, even though they could get kicked out if we looked at more concrete and specific instantiations of that trait. Given the never-ending diversity, all typical traits result from abstraction and thus from disjunctions. Only the respective scientific context can decide whether the abstracted property—the established abstract similarity between individuals—is useful (e.g., interesting, not ad hoc, carrying explanatory information, robust). The place for the stopper to the regress cannot be established once and for all.

All that can be said is that there needs to be a stopper and there needs to be some abstraction.

If there were no stopper, as Machery correctly pointed out, the concept of human nature would become too inclusive: every kind of variation would then in principle be included by constructing an abstract property out of the disjunction of the concrete variation. It is thus a justified worry that some empirical approaches seem to be very inclusively minded. Evolutionary psychologists, for instance, do not seem to hesitate to include a lot of things as part of human nature. If evidence on cultural variation comes in, they adapt their explanandum and go more abstract, so that there is again something that is shared across all human populations.

But if there were no abstraction, science would simply collapse. After all, any general statement requires some abstraction. Regarding abstraction in and of itself as illegitimate epistemic process would make sciences ultimately impossible. With respect to human nature, we would fail at the other extreme and become too exclusive. If we did not at all allow for inclusion via abstraction from a disjunction, then nothing would be part of human nature. Human nature would be an empty category. It would be empty not because of any interesting empirical results but because of ignoring the epistemology of difference making as contrasted with production. Although it might empirically well be that human nature is quite thin, it is not empty. There evidently are similarities as well as differences between humans, and pointing at them is relevant for different fields across sciences.

7.3 A Narrow Enough Concept of Human Nature in the Descriptive Sense

Channelism as the Foundation for the Solution

I want to take stock: it was important to clarify how something can be due to culture or due to nature because I accepted that habits like burying the dead are not part of human nature, even given that they are typical and evolved. It also helped to understand the role of abstraction: the kind of reconstitution of phenomena, the ignoring of certain differences (in cause or in effect) that simply are part of science.

The result is that a typical trait is part of human nature if the developmental resources that make a difference for the (abstracted) trait are conserved over evolutionary time by biological rather than cultural inheritance. Developmentally such a trait is nonetheless dependent on nature and culture: dependent not only on developmental resources inherited biologically but also on developmental resources that are culturally learned (or persist environmentally). Any trait is produced (i.e., develops) only when all causally relevant developmental resources are present.

This solution relies on the divide between different channels of transmission rather than on a divide that assumes that culture is not evolved or selected. The solution allows having traits that are due to culture or nature alone. But it does so only if one respects the distinction between making a difference (explaining a difference with respect to a trait) and causal production (explaining the production of a trait) and if one allows abstraction from a disjunction to be in principle epistemologically legitimate.

The Descriptive Nature in Review: Solved and Unsolved Issues

In comparison with Machery's solution, delineating different channels of inheritance allows for a descriptive human nature that has a clear contrast (biologically versus socially inherited resources) at the foundation of the etiological kind term *human nature*. It provides the narrowing needed (in addition to typicality) and is compatible with the interactionist consensus.

Given all these, it is the best descriptive nonessentialist successor concept since it is left untouched by the standard anti-essentialist critiques by being statistical and historical; it accounts for references to traits as being "part of human nature" and "due to nature"; it carves reality at one of its important joints: biological versus cultural inheritance. This descriptive concept of human nature (i.e., that there are stable generalizations about humans that are analogous to what one finds for other species) has, I think, empirically never been in doubt, even though its conceptual foundations have been philosophically contested.

A couple of aspects of that descriptive concept can be regarded as more or less settled. First, there is no species specificity assumed. Something can be part of human nature in the descriptive sense, even if other species exhibit the same trait (e.g., tool use). In sciences such as psychology or anthropology, this is a common usage except when the origin of *Homo sapiens* and the boundary between "them" and "us" is at issue, as still so often in philosophy.

Second, there is no universality assumed. Universality is replaced by typicality since the descriptive concept does not abstract an essence from the patterns of variations, the clusters observed within those identified as being part of the reference class (i.e., those genealogically belonging to the respective species whose nature is at issue). In short, there is no "all and only humans are/have/can exhibit trait X." There are only statistical clusters, and genealogy decides who counts.

Third, no evolutionary fixity is assumed. The statistical clusters that one can observe do not have to be innate in any strong sense: they change continuously even though they exhibit some stability. Development and evolution (catalyzed by the choices humans make) created human nature, changed it, and will continue to do so. There is no, and never was, an evolutionarily fixed set of characteristic traits of being human, even if there are reliably reoccurring clusters of properties for a certain time. Metaphorically, human nature in the descriptive sense is like a moving cloud. Literally, it is a cluster of properties that extends in space and time.

Fourth, the stability over time results from the fact that traits that are part of human nature rely on resources that travel the biological channel of transmission, which has a high stability built in because biological factors can rarely change from a vertical to another mode of transmission. Traits that are "due to culture" are traits that are typical or not, but for which the conditional (or respective counterfactual) holds: if they vary, the difference between them relies on a difference in developmental resources transmitted by cultural inheritance. "Speaking English" (rather than another concrete language) is a case in point.

Fifth, there is no normativity derivable from such a human nature. The biological normalcy that one can observe is statistical and evolutionarily contingent. It can thus not be used to develop a concept of normal (versus abnormal) functioning: everything that is viable at the level of the individual is "normal" from the biological point of view. As Amundson (2000) states, developmental plasticity secures viability even in face of statistically abnormal starting conditions (be they genetic or otherwise): "Development yields adults that *function*, but not adults that function *identically*," or normally (39, emphasis in original). This does not lead to an anything-goes situation. Recall the example of parents from chapter 5 who want to raise their child species-less (e.g., as a squirrel). There is a matter of fact that legitimates ethical intervention into the doings of the parents since it can be doubted that an entity born to humans can survive as a squirrel. Other ethical interventions (e.g., interventions regarding sexual orientation), though, do not have such a justification.

In contrast to these more or less settled issues, one issue is still in need of further in-depth analysis. First, if the descriptive concept itself does not fulfill an explanatory role, what else does? If there is a stable property cluster making up our human nature, there also has to be something that developmentally and evolutionarily explains it. If human nature in the descriptive sense is an etiological kind term, then it points to a specific kind of explanation. The discussion on stability and channelism in chapter 5 and in this chapter has already shown where it points to precisely: to the set of developmental resources that are reliably transmitted from generation to generation via biological inheritance. The next chapter develops this point further. The post-essentialist explanatory nature is a set of specific kind of causal factors: developmental resources that are typical and biologically inherited. Compared to the traits that are part of the descriptive nature, these causal factors—"nature" in the explanatory sense—are "due to nature" in a much more straightforward sense.

8   An Explanatory Nature

The descriptive nature I discussed in chapters 6 and 7 has an explanatory counterpart, an explanatory human nature. It is statements like "X is due to human nature" or "human nature causes X" that are at issue. An example is: "It is because of our human nature that most adult humans have a body height between 150 and 190 cm." Chapter 7 showed, among other things that whether individual examples of such human nature claims make sense is an empirical issue and thus depends on each case. Since the descriptive nature refers to a subset of typical traits of a species, the counterpart explanatory nature refers to a subset of reliably reoccurring developmental resources—to causes rather than to the traits that are reliably produced thereby.

This chapter defends the claim that when we use the term *human nature* in an explanatory sense, we should refer to the pool of developmental resources that travel via biological reproduction from one generation to the next. The reason is that it is the pool that is characterized by a characteristic stability. It is an explanatory nature that, strictly viewed, resides not in individuals but in populations.

In section 8.1, I reject the arguments from Walsh (2006) and Devitt (2008), who defend intrinsic essences in the explanatory sense, despite the Darwinian challenge. I then show in section 8.2 that the explanatory nature that can be defended belongs to a population rather than to individuals. In that sense, the account is not unlike some ideas from developmental systems theory, but it is less radical since I insist on the relevance of distinguishing different channels of inheritance for understanding the stability of the property cluster that I have called a descriptive nature of a species. In section 8.3, I summarize the completed picture about the three post-essentialist natures.

8.1 Explanatory Neo-Essentialism

In this section, I discuss why a purely explanatory neo-essentialism, as advocated by Walsh (2006) and Devitt (2008), is missing the point of population thinking with respect to the explanatory nature of a species. Both approaches defend what I call a typological individualism: they defend that there are explanatory essences that are—as traditional essences—intrinsic to individual organisms. I will argue that they are biased toward intrinsicality and have not taken the full force of a populational stance into account.

Recall that traditionally, the nature of a kind is regarded as consisting of its essence, which refers to something internal to the individuals partaking of a kind. These internals can be specified as intrinsic properties, causal capacities (or dispositions), constitutive structures, or developmental resources (depending on metaphysical account).[1] Whatever is chosen, traditional essentialism is individualist in addition to being monistic. Monism has been addressed in previous chapters. In its essentialist form, it states that an essence is simultaneously fulfilling a classificatory, an explanatory, and a descriptive role; in its nonessentialist form, it claims that only one of these epistemic roles survives the Darwinian challenge. Individualism in the context of the issues discussed here, by contrast, amounts to the claim that a nature is intrinsic to individual organisms. Even a pluralist can defend it.

This chapter thus sets a focus on what remains of the typological individualism implied in traditional essentialism, an individualism that is locating the explanatory nature of a species in individual organisms without taking their individuality seriously. The aim is to show that if intrinsicality is tied to what I call a typological individualism, this is the wrong road to take. I distinguish between two kinds of explanatory essentialisms: developmental and teleological. In the remainder of this section (unless otherwise noted), "intrinsic" is short for "intrinsic to individual organisms."

After the critique of developmental and teleological essentialism, I present my replacement: an explanatory nature of a species (be it humankind or any other animal species) that does not need any reference to a traditional essence and that means it has no need for a metaphysical argument toward a special status of intrinsic parts of organisms (Devitt), as well as no need for dispositional capacities of organisms (Walsh). It nonetheless uses the nature-culture divide and is thus more contrastive than alternative suggestions that take the Darwinian and the developmentalist challenge seriously.

Developmental Essentialism

Devitt (2008) defends a new "intrinsic essentialism," meant as a reply to the Darwinian challenge.[2] Species, according to him, "have essences, that are, at least partly, underlying intrinsic, mostly genetic, properties" (Devitt 2008, 344). This means that the explanatory essence of the species has to include developmental resources that are intrinsic to the bodies at issue. The account developed here also has the explanatory nature referring to developmental resources that are—as a matter of fact—internal to bodies. But the reason they are called a nature is, according to the account developed here, not because these developmental resources are internal or in any sense "essential"; they are called a nature simply because they travel a specific channel of inheritance and are therefore more stable. To see the difference between my account and Devitt's, I will reconstruct his approach in the language developed in the previous chapters.

Devitt's main argument is that (first premise) the partaking question can be answered only by a trait explanation. Trait explanations (second premise) require (partly) intrinsic essences, that is, intrinsic essential properties. Consequently, (conclusion) intrinsic essential properties are necessary to answer the partaking question. There is a variation of that argument that is independent of the partaking question. It considers the description question in relation to the trait explanation question. Since the latter two questions are more directly related, I will take this second version as the stronger one and thus as my main target. The argument then says that generalizations about a biological kind (e.g., that tigers have stripes) (first premise) make sense only on the basis of developmental trait explanations. So far, I agree. Such developmental trait explanations (second premise, as in the first version) require reference to intrinsic essential properties, that is, reference to developmental resources intrinsic to the organisms. Consequently (conclusion), intrinsic essential properties are necessary for making sense of generalizations about biological kinds.

The problematic part is in what goes unnoticed into the second premise and in what is further concluded from the argument. The second premise exhibits an intrinsicality bias, a bias toward intrinsic properties that conflicts with the interactionist consensus, since the same argument can be made by replacing "intrinsic essential properties" with "extrinsic essential properties." Whatever trait one looks at, it will necessarily be caused by intrinsic as well as extrinsic developmental resources. There is no development of traits (of specific organisms) from intrinsic nature alone, contrasted with extrinsic nurture, as described in chapter 4 under the label genetic inertness. Both kinds of developmental resources, intrinsic as well as extrinsic to the individual ones, are equally necessary. Thus, the argument cannot be the foundation of an explicitly intrinsic essentialism. If one asks for a developmental trait explanation, there is no reason that the explanation should give priority to intrinsic factors as more essential. Thus, the problem with the argument is not that it is not valid. The problem is what it leaves unsaid: that the exact same argument can be made for external developmental resources since they are equally necessary.

Devitt, however, assumes (implicitly and without justification) that intrinsic factors, qua being intrinsic, are essential factors, while extrinsic ones are not. Intrinsic factors are certainly contributing causes to the development of all traits that are typical of being human, but so do those extrinsic to the individuals (i.e., environmental causes). Why, then, not just say that developmental trait explanations require by necessity nature and nurture, intrinsic as well as extrinsic factors? If one restricts the set because one can always give only partial explanations, one should at least do that in an unbiased way or make the basis of prioritizing certain developmental resources explicit. My account sees the basis of prioritizing some of them in their stability; Devitt sees it in intrinsicality, which in my account is derivative only.

Devitt still describes the environment predominantly in the way criticized already by Sober's (1980) account of what is wrong with traditional explanatory essentialism (described in chapter 4): the environment is for him a mere trigger or a mere modifying or disturbing cause. Dewitt writes, for instance, "that the intrinsic essence of being F, together with the environment, explains the morphological, physiological, and behavioral properties typical of F's" (Devitt 2008, 374), with F referring to the species, e.g., *Homo sapiens*. Why are only intrinsic developmental resources partaking in the "essence" of F? The environment is just an environment? Never essential? But why? Devitt does not want to fall back to a gene centrism, with its notions of "coding," "information" and the like, which would make genes a special kind of developmental resource with a special causal importance.

An Explanatory Nature 173

Thus, although he acknowledges that genes are not special in that sense, he seems to believe that intrinsic causal factors, contrasted with extrinsic ones, are special nonetheless. But why? Simply by being intrinsic? When Devitt (2008, 378) explicitly answers why the intrinsic causal resources are "essential" (rather than just as explanatory as all the gory other developmental resources), he points to the way one explains why a field guide about the species F holds, that is, why a phrase "members of the species are X" holds (e.g., why "tigers have stripes" holds). Within his answer, he makes two wrong claims. First, Devitt sometimes seems to assume that only intrinsic properties are explanatorily relevant for such generalizations. This conflicts with the interactionist consensus and is clearly wrong. Second, he claims that the reason such generalizations hold is always explained by the same specific intrinsic property. At that point, his position conflicts with the fact of evolution since that intrinsic property might change via evolutionary change, without the generalization failing to hold, because of robustness or redundancy of the developmental system. Even when parts of the developmental resources change, the life form can stay the same. Biological systems are robust (i.e., multiply realizable). For instance, humans are said to need vitamin C for a healthy life. In a way, the need for it seems to be part of the descriptive human nature, at least from a medical-scientific point of view. But (if the scientific story about vitamin C is correct), we lost the ability to produce vitamin C internally since it became available in abundance environmentally, thanks to learning and niche construction (i.e., cultural evolution and changes in the environment). We learned how to get it from the environment in reliable and, at the same time, flexible ways. As a result of a long coevolutionary process, our explanatory nature (for the former ability to produce it internally) got externalized. Consequently, the ability to produce vitamin C is not part of our descriptive nature anymore, even though the need for it still is.

To conclude, against Devitt, I claim that it is not necessarily because of the same properties internal to the organisms that a specific generalization about humans holds. Generalizations about humans contingently hold (in space and time) because of the sum of all developmental resources that are typical and stably reoccur in the population of humans. The internal resources are "nature" rather than "nurture" not because they are internal, but because they are more reliably reoccurring.

I am thus not denying that human nature in the explanatory sense refers to developmental resources intrinsic to individuals. I am only denying that this is necessarily so. The reason human nature is intrinsic is quite contingent: multicellular organisms evolved mechanisms of biological inheritance that make it very likely that developmental resources that travel the reproductive channel reoccur stably across generations (certainly only as long as selection supports it). Developmental resources intrinsic to individuals therefore convey a stability that is necessary for there to be a descriptive nature, but neither are the factors accounting for stability necessarily intrinsic nor is a specific trait by necessity always due to the same set of intrinsic developmental resources (as the example of vitamin C illustrates).

Thus, the explanatory nature one can point to, in order to stay consistent with contemporary developmental and evolutionary biology, does not exclude that there are intrinsic factors, but it does not give them priority by necessity or intrinsicality. There is a priority, but it is a gradual and contingent one. Only because the biological channel is making a high evolutionary scope of a developmental resource much more likely than the cultural channel is it regarded as the nature of being human in the explanatory sense: it is what is virtually given, since for us, for each myopic individual organism, it appears stable. As mentioned at the end of chapter 4, reference to an internal-external divide to make internal developmental resources special without any further argument is a sign for a metaphysical intrinsicality bias that should be prevented.

I am thus arguing against a metaphysical intrinsicality bias, not against intrinsic properties (e.g., genes, epigenetic, and other cellular factors) as being the most reliable part of the cluster of developmental resources necessary to make those surface clusters of properties possible that describe a species in the sense of a descriptive human nature.

Teleological Essentialism
Walsh (2006) accepts the first part of the Darwinian challenge: that talk about a definitional essence is pointless from the perspective of modern biology, yet as he argues, this does not exclude that there is something like an explanatory essence, which in his account amounts to those capacities of organisms of a kind that explain evolvability of the kind. Evolvability in turn requires robustness, developmental plasticity, and adaptiveness. I

agree with the latter, but not with the kind of individualism inherent in his account. Like Devitt, he is subject to an intrinsicality bias.

Walsh takes evolvability of the kind as the major phenomenon that is in need of explanation. This requires that individual organisms of a kind exhibit plasticity (in conjunction with robustness and adaptiveness). Walsh implicitly also assumes that the nature of a kind reduces to the natures of individual organisms of the kind—with their species-specific and typical capacities for plasticity and the respective intrinsic resources for development underlying the respective individual capacity for plasticity. Although Walsh talks about individual natures, each organism of a kind has, according to him, basically the same formal nature in the sense of a species-specific and typical capacity for plasticity. And since it is a capacity, it can be realized in different ways, depending on environmental variables.

With this, he aims to defend a teleological essentialism that is, according to him, Aristotelian in the following sense: plasticity for Walsh is an Aristotelian formal nature that is connected with the telos of a kind. The respective developmental resources are then corresponding to the Aristotelian material nature only. Not only is there a formal nature, but also the kind of explanation it provides is Aristotelian, in the sense that it is teleological rather than causal. When Walsh (2006, 432) states that "natures of organisms explain their salient features and that shared natures explain resemblances among organisms," then explanation is not causal but teleological.[3] It has the relation between natures and generalizations (in this case, explanatory generalizations) just upside down compared to Machery and me. In an Aristotelian picture, natures teleologically explain why there can be generalizations. In a post-essentialist picture, generalizations "explain" (i.e., make understandable) what people mean with the term *nature*.

Walsh's account is subject to an intrinsicality bias similar to the one in Devitt. He gives priority to intrinsic developmental resources (understood as an Aristotelian material nature) because he postulates a capacity for plasticity in individual organisms that he equates with the formal nature of these individuals. It would be equally possible to say that the capacity needs extrinsic developmental resources. Intrinsic traits of individuals are not the only resources that are important to explain why species in general or any particular species is evolvable. Intrinsic as well as extrinsic developmental resources will be important for plasticity (and ultimately for evolvability). As I mentioned in chapter 4, a whole bunch of traditions from the twentieth century (e.g., dialectical biology and developmental systems theory, or evo-devo) have stressed this, in part against genetic reductionists believing that a specific subset of intrinsic factors, genes (versus all the rest), explains it all.

Walsh does not explicitly answer Sober's and Hull's arguments against variation-reducing capacities that I discussed in chapter 3. This is a pity since it seems that he has nothing on offer to reply to Sober. Given the populational way of accounting for variation, as discussed in detail in chapter 3, one does not need capacity talk to have an explanation for why things are the way they are. If all developmental resources for a specific trait of organisms were mentioned (i.e., if the "lazy" ceteris paribus condition on when the respective capacity is realized, or prevented from being realized by disturbing factors, would be spelled out), then there would be no place for the essential capacity of tigers to develop stripes.[4] As Sober said, in a populational picture, there is no longer a distinction between a natural state (realized given the right ceteris paribus conditions) and a nonnatural state (due to disturbing causal factors). Most tigers develop stripes, some do not, and for each individual tiger, there is the same kind of explanation: a developmental trait explanation. One can use an elliptic capacity talk (and the respective ceteris paribus talk or generics talk), but to do so does not make what one invents thereby, namely the capacity, explanatory. According to this, the formal nature Walsh appeals to is an abstraction only. For descriptive purposes, these are legitimate, but not for explanatory purposes.

Each individual has what it has and develops according to the actual developmental resources it has: its individual genes and its environmental resources: no range, no norm, no possibilities, no disposition to develop otherwise. An individual does not have a capacity for a reaction norm or a "phenotypic repertoire" (Walsh), for example, for body height, in the sense of the capacity to become between 150 and 190 cm tall, even if that is the range of body height typical of humankind. Granted, these capacities are useful abstractions: one abstracts away from the variation, that is, how tall exactly the respective individuals are, as one abstracts away from the differences in concrete language when one speaks about humans having a language. One can thus describe the type of individuals in an abstracting manner. Abstractions are a useful way to talk in the sense of Dennett's (1991) real patterns (i.e., patterns at the populational level), but they are not in reality existing in individuals and they are not explanatory. At least Walsh (2006) does not offer the resources to convince one otherwise. As abstractions, capacities for this or that (i.e., phenotypic repertoires in Walsh's language) are statistical properties of a population and therefore should rather be called population-level kind propensities.

Not all human beings speak, and those who do speak, do so differently. What is typical is that humans have *a* language, despite the empirically observable differences. Does that mean that there is an ability for human language in every individual? Is a typical trait realized as individual disposition in all humans? According to Devitt and Walsh, there is. According to the populational picture defended here, it is not. What is typical is an abstract property, the ability to have a language, exhibited by a majority of people. Adults who are unable to communicate using language at all simply do not have a capacity to develop it. Only those who use a language have had that capacity to develop it.

I am thus not denying that the trait of having a language exists. I am only denying that each human intrinsically shares the same capacity qua being human. Whatever kind of property it is, there will be variation with respect to that property between individuals and no capacity talk is needed to predict that variation or to explain why there is a high probability that the next individual of that kind will equally be between 150 and 190 cm tall or communicate the way other *H. sapiens* do. As Sober reconstructed the populational way of explaining variation, previous variation, that is, the pool of developmental resources available for a specific population of individuals, predicts and explains it, with the biological inherited pool having a higher scope because it makes the reoccurrence of developmental resources more likely and is in that sense more stable.

In an Aristotelian picture like the one Walsh assumes, an explanatory nature explains why there can be generalizations and thus kinds. In a post-essentialist picture, this is not just turned upside down but chopped into a pluralist populational picture: the genealogical nexus (classificatory nature) constitutes a channel of transmission (biological heredity by biological reproduction). It in part and indirectly explains why there is a pool of developmental resources within a population that reliably reoccurs over time, independent of the choices of individuals for quick adaptation through culture. This in turn explains (in part, but importantly so) why there can be generalizations about that population at the phenotypic level. Genealogy is thus involved in the explanation of why there can be an explanatory nature and a descriptive nature. The pool of developmental resources that make up the explanatory nature can be reified and individualized by capacity talk and called explanatory in a causal or teleological sense, as if there were a capacity. But to do so is not explanatory. Thus, if parsimony is of epistemic value and explanation is at issue, then the populational account should get the support and any typological individualism should be rejected.

There is one way to reconcile Walsh's explanatory essentialism with the Darwinian challenge. The entity in the world that actually has a reaction norm (or range of reactions), and thus the possibilities or capacities at issue, might be specified as the population rather than the individuals themselves. Thus, if one were to use capacity talk (which one can as an epistemic short-cut), then the respective capacities are actually a property of the population, not of the individual organisms, and should be called by a different name, for instance, *propensities*. If one chooses that way of talking, there are no shared dispositional essences intrinsic to individuals, but there is a pool of biologically inherited developmental resources intrinsic to populations and metapopulations (i.e., species). These populational repertoires (propensities, if you like) partly constitute a range of reaction for individuals. Thus, one can have something like an explanatory essence in the Aristotelian sense, a formal nature (propensity) with a material nature, but the bearer of that material and formal nature has to be rather un-Aristotelian, anti-individualistic, to reconcile having a nature in that sense with the second anti-essentialist claim of the Darwinian challenge, the essential variability of all developmental resources.

To recap, a teleological essentialism with its reliance on capacities (Walsh), as well as a developmental essentialism with a metaphysical bias toward intrinsicality (Devitt), should be rejected on the same grounds as traditional essentialism has been rejected by authors like Hull and Sober: because they rely on assumptions (intrinsicality bias, capacity talk) that are superfluous for contemporary biology and often even incompatible with it. Since there is an alternative, these essentialisms are not just ignoring variation; they are also mistreating variation and its sources. Furthermore, the typological individualism that is part and parcel of Walsh's teleological essentialism (i.e., the idea that each individual organism of a species has the same kind of capacity) should be replaced by a populationist individualism in order to really appreciate individuality (i.e., variation).

8.2 A Population-Level Solution

I agree with Devitt that the thing that performs the explanatory role of an explanatory essence (the answer to the developmental trait explanation question) contains developmental resources intrinsic to the individual. Yet only the whole pool of developmental factors across individuals, transported from generation to generation by biological reproduction, can be the explanatory nature of a species—what one can point to in order to explain why the generalizations that are made about species members hold (why the respective properties are typical and stable).

Explanatory Nature as Intrinsic to the Population

In replying to Devitt and Walsh, I already introduced the alternative: the explanatory nature constituted by the mechanisms that make biological inheritance possible and stable is populational; it is the pool of developmental resources intrinsic to the population rather than to individuals, since no individual shares the exact same developmental resources. Taking this into account is necessary for taking the Darwinian, the developmentalist, and the dehumanization challenge seriously.

I now consider an important reductionist reply. Boulter (2012), in a defense of a developmental essentialism similar to the one from Devitt, claims that the properties of populations depend on properties of individuals, so that the populational solution could be reduced to an individual-level solution. He claims that "Sober's argument that the properties of individuals can be ignored in population thinking" ignores

> that the statistical properties of populations are ontologically dependent upon the properties of the individuals that make up the population. So at some explanatory stage the properties of individuals must be factored in. Their essential properties will be among those adverted to in the course of this level of explanation. (Boulter 2012, 102–103)

The crucial point in my reply is to keep ontology and causality apart. Ontologically, Boulter has a point, and that is precisely why variation is so important: individuality grounds populational thinking. But even ontologically it holds that individuals, strictly speaking, do not share an explanatory nature, as they do not share the descriptive nature, strictly speaking. After all, the problem of squaring the circles, discussed in chapter 3, also applies to developmental resources. And it is solved in the same way:

human nature in the explanatory sense is a set of developmental resources that is typical and stable, reoccurring for the species in space and time. We can thus ontologically take the following as the explanatory counterpart to the descriptive nature:

> Explanatory nature = causal factors (developmental resources) that are typical and that are due to biological inheritance.

If so, then causally it holds that the explanatory stage that Boulter is talking about is itself only a result of a previous explanatory stage of the population at time t_0, making it more likely that an individual at time t_1 has the individual properties it has. What is causally prior? The population or the individual? If that circle of trait explanation is all one looks at, then it is hard to say. But since genealogy is providing a deeper answer since it is involved in answering all five questions I set out in chapter 5, I opt for an explanatory priority of the population. Consequently, the resulting populational pool of developmental resources itself is the bearer of the explanatory nature (not the individuals who house these resources temporarily).

Galton as a Population Thinker

Galton himself had a tendency (with a romantic bent it seems) to think in that direction:

We must not permit ourselves to consider each human or other personality as something supernaturally added to the stock of nature, but rather as a segregation of what already existed, under a new shape, and as a regular consequence of previous conditions. ... We may look upon each individual as something not wholly detached from its parent source,—as a wave that has been lifted and shaped by normal conditions in an unknown, illimitable ocean. There is decidedly a solidarity as well as a separateness in all human, and probably in all lives whatsoever. ... It points to the conclusion that all life is single in its essence, but various, ever varying, and interactive in its manifestations. (Galton 1892, 376)

Population thinking not only teaches one that the pattern of evolution is variational rather than transformational, but also that the bearer of the explanatory essence, that bundle of especially reliable reoccurring resources for the similarity of the members of the kind, exceeds the organism boundary. It is a pool of developmental resources, an ocean with "individual waves" (to use Galton's metaphorical words) partaking in it.

Comparison to Developmental Systems Theory and Samuels

Griffiths (1999, 2011) and Stotz (2010) can be interpreted as allowing for something similar: that a genealogical approach, combined with the homeostatic property cluster approach, results in a concept of a relational human nature that comprises all the resources needed to stabilize the development of the patterns of similarity and difference observable in humankind. Human nature in the explanatory sense, the thing that explains how humans are, is then a genealogically anchored explanatory essence of gigantic size: the whole developmental system of humankind, including the developmental niche (Stotz 2010).

The main difference from my account is slight but crucial. Griffiths and Stotz are, first, too inclusive with respect to the descriptive nature and, second, too monistic in general outlook. First, when everything in the pattern of similarity and difference is part of human nature, the claim that there is a human nature is not just becoming too inclusive, as mentioned earlier with respect to the regress problem regarding inclusion of polymorphisms, but indistinguishable from a claim (e.g., the existentialists) that there is *no human nature*. Stotz and Griffiths are too inclusive as the existentialists are too exclusive: they do not allow us to make empirically informative, and that means contrastive, claims: to say, for instance, that carrying a cell phone is not part of human nature, while acknowledging that having a language clearly is. Patterns of difference and patterns of (in-)dependence become invisible as part of the "all-included" developmental systems concept of human nature, as they become invisible in human nature denials. As Levins and Lewontin (1985, 256) already stated with respect to the existentialist's point of view, such a view "leaves us with no way to understand human society"; from a too exclusive or too inclusive perspective, human nature and human society "simply is what it is." Second, Stotz and Griffiths fail to acknowledge *natures in the plural*, even though the developmental systems approach could in principle allow for it by acknowledging that, first, some scientists might not be interested in explaining why we are the way we are (being interested in a descriptive concept of human nature only), and that, second, some might not be interested in the specific project of developmental systems theory, namely, the explanation of development. Like other accounts, Stotz and Griffiths make a choice—a choice of focus, of epistemic interest. They, as anybody else who tries to capture the complexity of life, focus on certain issues and leave out other things. In their case, they leave out, for instance, the temporal order in heredity, the different kinds of stability inherent in the different channels of inheritance, and also the autonomy of culture. They are not using the channelism defended here and consequently ignore the dynamic difference between biological and cultural inheritance. For them, the complete developmental system is human nature. Full stop. Such an all-inclusive concept, as already mentioned in chapter 6, fails to have contrastive power: interpreted as a descriptive concept, it is indistinguishable from a claim that denies that there is any descriptive human nature; interpreted as an explanatory concept, it fails to distinguish between dynamically different bundles of developmental resources. They ended up with such an all-inclusive position since they take the Darwinian and the developmentalist challenge seriously, and that is good. Nonetheless, they went one step too far by denying or leaving out the importance of channelism advocated here. Stotz and Griffiths (forthcoming) reply to the claim that their account is too inclusive, an issue already raised with less detail in Kronfeldner et al. (2014). The developmental systems theory account of human nature is not useless, they claim, since it is at least better than the essentialist account and development simply is complex. Both claims are true, but it does not answer the critique against their all-inclusiveness since the critique here and in Kronfeldner et al (2014) is not (as in standard critiques against developmental systems theory) that complexity is taken into account (i.e., that developmental systems theory cannot study all causal factors simultaneously) or that the environment enters the picture. The critique is that, first, the developmental systems theory account is too inclusive (with respect to the descriptive as well as the explanatory nature of humans) and that, second, it is on a par with other accounts in being epistemically selective with respect to their core explananda.

Samuels's (2012) explanatory account is closest to what I intend here. According to him, the term *human nature* should be taken to refer to an explanatory nature. He defines it as "a suite of mechanisms that underlie the manifestation of species-typical cognitive and behavioral regularities" (2012, 2). These mechanisms are themselves regularities (e.g., a language production system common in humans), but regularities that explain other regularities, namely, surface patterns of similarity and difference (e.g., similarities and differences in spoken or signed languages).

Like Griffiths, Stotz, Ramsey, Lewens, and Machery, Samuels is not taking the pluralist stance that I defend here, even though he admits, as some other monists do, that his successor concept is not fulfilling all three scientific roles. He votes for one specific successor concept and claims that it fulfills a larger set of epistemic roles than Machery's nomological alternative.

Furthermore, Samuels (2012, 23) explicitly claims that his account is not assuming "that essences must be intrinsic." On the contrary, he regards the whole suite of homeostatic mechanisms that I discussed in chapter 5, external and internal, as equally relevant for explaining why properties cluster. Samuels (2012, 23) adds that the homeostatic mechanisms can be distinguished according to the "timescales" they operate: there are "evolutionary mechanisms" (such as selection or drift) that "cause human species-typical properties"; there are "developmental mechanisms" that are "responsible for the acquisition of human psychological capacities" (such as "those involved in the development of the neural tube," or mechanisms of "conditioning, induction and other sorts of learning"); and there are "synchronic mechanisms" (such as visual processing mechanisms) that "are causally responsible for particular manifestations of psychological capacities." He adds that "if we aim to provide *comprehensive explanations* of species-typical regularities, then all of them are presumably relevant" (emphasis added).

Although his approach is very close to mine, there are some important differences. His approach ends up being too inclusive (as with Griffiths, Stotz, Ramsey, and Lewens) since it includes all mechanisms that are relevant to explain everything that humans do, including the "particular manifestations of psychological capacities" (e.g., speaking Japanese rather than English). If everything is connected to everything and is called explanatory human nature, then there certainly is such an explanatory human nature but in such a noncontrastive sense that it will be (again) of no use for sciences.

That is probably why Samuels also mentions that scientists will select among the mechanisms (and phenomena) they study and restrict the term *human nature* pragmatically to a subset of these. He claims that cognitive scientists will focus on mechanisms that he calls developmental and synchronic. Human nature for the cognitive scientist is then limited to the "suite of empirically discoverable proximal mechanisms—a causal essence—that causally explains the various psychological regularities that comprise our nomological nature" (Samuels 2012, 24). That means that cognitive scientists ignore certain causes and treat others as explanatory human nature—the relevant causes to explain what they want to explain. Samuels does not say so, but I take it to follow that biologists or evolutionary psychologists would go for a different set of causes, namely the evolutionary mechanisms, the so-called ultimate causes, and regard them as the respective causal essence.

His pluralism (decisively distinct from mine) faces some problems. First, Samuels uses the much criticized ultimate-proximate distinction, briefly discussed in chapters 5 and 7. Second, Samuels does not distinguish between the constitution question and the trait explanation question. Consequently, he takes "phylogenetic processes and mechanisms that operate over evolutionary time," such as selection, or niche construction, or interbreeding, or genealogy in and of itself, as explanatory in the same way as the products of these processes, the evolved psychological mechanisms. And even if he distinguished the two questions, it is likely that he would still regard the whole suite of homeostatic mechanisms as equally part of human nature, ignoring thereby not only the distinction between constitution explanations and trait explanations but also (and much more important for the account defended here) the distinction between different channels of transmission (even though he is not explicitly denying them, in contrast to developmental systems theorists). He thus seems to end up with an equally all-inclusive position similar to Griffiths, Stotz, Lewens, and Ramsey just at the explanatory level.

He seems not to see the importance of the nature–culture divide since he does not discuss whether different kinds of homeostatic mechanisms convey different degrees of what he calls "counterfactual robustness" and I stability.

8.3 The Explanatory Nature Established

The explanatory nature I defend is a pool of developmental resources that is intergenerationally transmitted by biological inheritance. The distinction from developmental resources that travel via cultural transmission points at something real: it refers to causal factors that are important in our explanation of development as well as evolution. The stability that the explanatory nature has and its distinctness, however, do not justify any christening of the set of causes as "explanatory essence" contrasted with external-to-the-individual developmental resources. It is an explanatory nature contrasted with culture because it gives much more stability, not because it is more essential in the sense of a traditional hidden, intrinsic, and unobservable essence.

Since even for this explanatory nature, variability is key, there is no universality of causal factors; typicality is all one will get. Let me thus adapt Hull's minimal concept of descriptive human nature to arrive at my final formulation for specifying the explanatory nature of being human:

> Explanatory human nature = statistical cluster of biologically inherited developmental resources that happen to be prevalent and stable over a considerable time in the evolutionary history of the human species.

No Conflict with the Developmentalist Challenge

The distinction between different channels of transmission is the only space between nature and nurture that is not a "mirage," to use Keller's (2010) term again. The explanatory nature, the pool of developmental resources handed down via biological inheritance, exists, even though the traits that are typical of the human species are always produced by all the factors from all the channels of inheritance (i.e., by nature, culture, and ecology). The account is thus compatible with taking further causal factors seriously, evolutionary and developmentally, as required by the interactionist consensus. The account refers to an explanatory nature as real, since the explanatory nature exists despite the interaction of these factors, broad or narrow sense, developmental, intergenerational or evolutionary.

The "Slate" Has Become a "Tube"

The famous slate is then pretty full, as depicted in figure 8.1, and is actually a tube (to imagine that, you have to look at the figure as three-dimensional).

The explanatory nature is a historically and statistically individuated entity: human nature in this explanatory sense exists as a property of a population that changes over evolutionary time. The organism is not the system that bears the alleged explanatory nature. One has to move to a populational level to make sense of a species' nature, be it in the classificatory, descriptive, or explanatory sense.

The explanatory nature defended here fulfills the explanatory role without importing debris from the essentialist picture (such as monism, an intrinsicality bias, or normalizing capacities that feed into dehumanization)

Figure 8.1
Human nature as a full tube. If one could literally see the explanatory human nature in action, it would look like a dense network of causal interactions with a frame. It would be near-decomposable as a subsystem of a bigger whole, like a tube extending in time and space (rather than a slate, as in the traditional metaphor of the blank slate), part of a bigger network of interaction. By seeing that in a certain area there is more interaction within an area than between that area and another, we draw a boundary and delineate inside and outside the channel. Since there are also interactions between inside and outside, the channel is characterized by porosity, that is, some lines continue toward the outside. (Illustration: Vera Brüggemann, Bielefeld.)

and without losing contrastive force (given that the essentialist picture had quite some contrastive force).

Furthermore, the explanatory nature is not necessarily a species-specific one since developmental factors do not need to be specific for a species in order to be scientifically useful for explaining the human life form. As noted, the epistemic roles of classification and explanation need to be decoupled. Uniqueness (or "unicity" as some call it) is important for classification, dividing between kinds, but not for explanation of typical and stable properties of a kind, except it feeds into an additional qualifier: what—in a given context—is regarded as important.

The Connection to the Classificatory and Descriptive Nature and the Division of Explanatory Labor

Despite the decoupling of epistemic roles in the pluralist picture defended here, there is a connection between the three natures of a species provided by genealogy. Genealogy constitutes biological heredity, that is, a channel of transmission of causal factors. Through it, the individual not only gets the status of being human but participates in the pool of biologically inherited causal factors typical of the species. The biologically inherited developmental resources are explanatorily distinct and of special importance because of their immense stability.

Last, but far from least, the pluralism implied here is inscribed in the division of labor of contemporary life sciences. The epistemic role of the explanatory nature is not to bring all causal factors of phenogenesis together onto one slate (or tube) in order to treat them on a par; on the contrary, because it carves reality at one of its joints, namely, at the joint of two clearly distinguishable subsystems of inheritance, it distinguishes different kinds of causes. It groups them into different kinds, with experts responsible to study those different kinds. The epistemic role of explanation has thus attached to it a pragmatic function of demarcating styles of inquiry along the lines of kinds of causes studied by these experts (Kronfeldner, forthcoming), establishing thereby an explanatory division of labor, with different experts studying different kinds of causes. If scientists are not interested in providing knowledge about the explanatory human nature, it can be treated as a disciplinary primitive (i.e., as things assumed but not studied). There is a right to ignore human nature. This holds especially if differences (between individuals or populations, in time or space) rather than similarities are of interest in the respective scientific field, as, for instance, in cognitive difference research.

In the next chapter, I will analyze how the pragmatics of ignoring certain causal factors in an explanation is ultimately involved in creating what it first aims to explain. We constantly (re-)create human nature in the explanatory and descriptive senses. Both the explanatory and the descriptive nature are real, yet they are in part made by us—not just constructed in our minds, but literally created in reality.

9 Causal Selection and How Human Nature Is Thereby Made

If our epistemic goal is to explain certain phenomena but there are too many causal factors involved, then one can ignore these factors by making some causes explanatorily irrelevant. This is achieved by adapting the explanandum, that is, by reconstituting the phenomena to be explained so that the resulting phenomena can be, for instance, as much as possible due to nature (or due to culture, depending on case). This abstraction method has been discussed in chapters 6 and 7. But there is another, and even simpler, way to ignore causal factors: leaving them out of the picture by selecting only some causes for consideration. Explanations are, after all, rarely complete. Scientists often start with a preferred causal factor, such as genetic resources for development, and proceed from there. In this chapter, I do two things: analyze why certain scientists (or people more generally) have preferences for specific kinds of causal factors and describe how human nature is literally (not just symbolically) made thereby.[1]

Within philosophy of science, the question about preferences for specific kinds of causal factors is dealt with under the label "causal selection." I will combine this debate, especially with respect to the theories from Collingwood (1940) and Hart and Honoré ([1959] 1985), with what I call explanatory looping effects, an idea analogous to Hacking's (1986, 1995, 2007a) classificatory looping effects. Explanatory looping effects will show how human nature can be changed in reaction to explanations.

The two main claims are that, first, a willingness-to-control principle accounts for many cases of causal selection and, second, people act and react differently, depending on the explanations for phenomena they adopt for an issue, which leads to explanatory looping effects. Ultimately these different reactions stabilize or change these same people, or their descendants, and thereby they change human nature in the explanatory sense (i.e., with respect to the distribution of the developmental resources) as well as in the descriptive sense (i.e., with respect to the distribution of traits characteristic of humankind). What we choose to see as causes sometimes makes who we are. In other words, human nature sometimes changes as a result of the choices made epistemically. Since this involves an important reflexivity of humans making humans via explanation, I chose the label "explanatory looping effect" for such processes.

In section 9.1, I introduce the debate and show how Collingwood's controllability account of causal selection can accommodate the claims made by others (e.g., Hart and Honoré and more contemporary revisions of their so-called normality account of causal selection). Yet Collingwood faces some problems that will be dealt with in detail in section 9.2 in order to establish the first main claim of this chapter: it is the willingness to control, not the ability to control, that often biases us toward certain causes and not others. By going beyond philosophical toy examples in section 9.3, I apply the issue to ignoring nature or culture in our explanations. Finally, the existence of explanatory looping effects is established in section 9.4.

9.1 Causal Selection, Control, and Normality

Partial Explanations
There are always many causes involved in the coming into being of something. For instance, many factors are causally involved in the etiology of body height, even if each cause has only a small influence. In principle, they all have to be taken into account to get a complete causal explanation of how body height is produced.[2]

In fact, complete causal explanations are rather hard to get, or too expensive, if not impossible, because they are completely beyond our scientific abilities.[3] Thus, when one explains a phenotypic trait, one first establishes a causal structure: a representation of a few candidate factors, which are likely to be causally relevant given prior knowledge. One includes, for instance, one genetic factor and one environmental factor in a norm-of-reaction graph and leaves out many other causal factors. The latter are either ignored, because they are unknown, or fixed at a certain value or randomized because they are not of interest. In any case, they are in the background. They are not looked at; they are like threads in a Jackson Pollock picture that are cut off by a frame, as in figure 8.1, and are therefore not perceived. In practice, scientists thus produce only partial explanations.[4]

As long as the partial explanations all pick out causes that are ontologically on a par, causes in the same sense (e.g., difference makers), they will all have in principle an equal right for being called correct, even if they all pick out different causes and thus give alternative explanations. Those explaining the difference in body height between men and women with reference to biologically inherited developmental resources are equally correct as those explaining the change in averaged height between the fifteenth and the twenty-first centuries with reference to culturally or environmentally inherited developmental resources. One can also take an example that does not involve different differences as explananda, for instance, "The cause of malaria is the bite of a mosquito." This is a partial causal explanation, since one also needs, for instance, a body that can be affected by the bite. The body is a cause too. But to make such a monocausal claim that leaves out the body is a perfectly fine causal explanation, despite its partiality, as Collingwood (1940, 299) stated in his analysis of partial causal explanations.

Principles of Causal Selection

How is it decided which causal factors can legitimately be ignored? According to John Stuart Mill (1858), who already addressed the issue, one selects causes in a capricious manner. D. Lewis (1973, 558–559), in the same vein, wrote, "We may select the abnormal or extraordinary causes, or those under human control, or those we deem good or bad, or just those we want to talk about. I have nothing to say about these principles of invidious discrimination." Lewis's aim was to understand what he called a "nondiscriminatory concept of causation." He wanted to explicate what a cause *is*. Causal selection points to something else, an epistemological rather than a conceptual problem, and addresses why people select certain causes and not others. That there are pragmatic aspects involved, as Mill and Lewis rightfully pointed out, does not exclude that there are widely shared principles of causal selection that help to understand why scientists are doing what they do when they give some causes priority over others. In addition, the two problems (the conceptual problem Lewis was interested in and the epistemological problem of causal selection) are independent. Thus, the account of causal selection developed here, will be independent of any specific theory about the concept of causation. I will nonetheless continue to use the language of difference making. Causes are difference makers, without deciding between the many theories of difference making and the metaphysics of causation. The so-called problem of causal selection can thus be put as the question of whether there are shared, general principles that guide us (and maybe legitimately so) in our discriminatory way of dealing with causes that are ontologically on a par.

Collingwood famously proposed a principle of causal selection for the context of practical sciences—knowledge domains such as law, medicine, or technology in which things getting wrong are studied. He claimed that in such contexts, if I am the explainer, the selected cause is "the thing that I can *put right*" (Collingwood 1940, 303, emphasis added). The explanatorily relevant cause is the thing that one can control, in the sense of putting it right.

One of his toy examples was the following. Imagine that one is driving a car uphill, and it stops in the middle of the hill. Why does the car fail to climb the hill? As he mentions, the hill is a perfect cause since "more power is needed to take a car uphill than to take it along the level." Whether there is a hill or not makes a difference, and thus the hill is causally relevant for the stopped car. People nonetheless ignore the hill as a cause, and rightfully so, he added, since people usually cannot intervene on hills (e.g., by stamping on them). One rather quotes as cause the loose high-tension cable, which (let's imagine) one finds after investigating the innards of the car. Collingwood's claim is that it would be pointless to say the hill is a cause of the stoppage of the car, even though ontologically viewed, it is a cause on a par with the loose high-tension cable. What guides us in causal explanation according to Collingwood, and justifiably so, is whether we can intervene or not—in other words, controllability.

Hart and Honoré ([1959] 1985) argued that it is normality (in the sense of frequency) that guides us in causal selection. We select the cause that is anormal (i.e., unusual), and what is normal (i.e., usual) is ignored and treated as background condition. Normality, they argued, is the more inclusive principle. Before I show that it is instead the other way around—that control is more inclusive than normality—I will spell out their account in a bit more detail, because it is important to see that normality already involves a choice—a choice of reference class, which goes back to the problem of extrapolation discussed in previous chapters.

The Causal Field and How Normality Is Chosen

In the terminology of Mackie (1974), when one causally explains something, one assumes a specific causal field.[5] In the malaria example that I mentioned, the mosquito bite is a cause only given a certain causal field. Only if one excludes entire causal settings (rather than just individual causes in a setting), by choosing the reference class accordingly, is a mosquito bite reliably (i.e., regularly) a cause. Some people are naturally immune to malaria. If they are part of the causal field, a mosquito bite is not reliably making a difference to whether an individual gets malaria. Without also intervening on their immunity, the people will not get malaria even if a mosquito carrying the disease agent bites them. When one says that mosquito bites are causing malaria, one assumes, implicitly or explicitly, that statistically normal people are not immune to it—that immunity is infrequent, not normal. For instance, when "we" (let's say tourists traveling to Africa) talk about the causal explanation of malaria, we often exclude "them" (those with a natural immunity) from the causal field simply because "we" are not "them." Some other people, however, especially those investigating or curing malaria on a professional basis, will have included that group more easily than tourists. In a similar sense, if the causal field in explanations of body height is defined by the nutritional heavens (as mentioned in chapter 4 where I discussed the problem of extrapolation), that means, if the people studied, the reference class, are exclusively from affluent backgrounds, then nutritional differences will not make a difference to body height. If, however, a "metabolic ghetto" is compared with a "nutritional heaven" (as described in studies on health inequalities, such as Wells 2010), then nutrition is clearly making a difference to health. The choice of worlds, including the choice of possible worlds, is decisive for what makes a difference. The choice of worlds (i.e., causal fields) equals the choice of reference class. These choices have a normality resulting. What is normal and abnormal can thus itself be picked out, a point that I take from Gannett (1999).

The presence of "normal people," who are immune to malaria or are not subject to metabolic ghettos, is analogous to the assumption that oxygen is normally present when we explain why a fire occurred (for a particular case or for fires generally). For instance, when we explain why a nearby barn burned down the other day by saying that the fire was caused by kids playing around with matches, we put the presence of oxygen in the background. In the context that we normally assume for this example, oxygen is a background factor for the fire, because in that context, oxygen is ubiquitous, and thus normal.

The analogy to this stock example of philosophical analysis on causal explanation illustrates that via choice of reference class, one gets normality (in the sense of frequency), and that influences what is considered as a candidate cause and what is left out of the picture, set fixed to be lingering around in the background as so-called standing condition. Hart and Honoré ([1959] 1985) have become the seminal reference for this "ab/normalism."[6] It is the abnormal, the thing that differs (e.g., whether there was a lighted match) that "makes the difference" (a language they used) in phenomena such as accidents, diseases, and deviant behaviors, that is, things defined as not "going on as usual" (Hart and Honoré [1959] 1985, 35). For the fire, an abnormal event, the presence of oxygen is not a cause (i.e., a difference maker) because oxygen is present when a house burns (unusual event) and when it does not burn (things going on as usual). But the lighted match is not normal in the context at issue. Thus, it is the cause.

If the normality criterion is interpreted that way, then it is similar to the contemporary account from Waters (2007), who uses a different language and drops the requirement that the explanandum itself has to be abnormal.[7] In the usual cases of fire, the case that one usually has in mind when we hear the example, presence of oxygen is—according to Waters—not an "actual difference maker" of a fire because in these cases, oxygen does not actually differ; in other words, it is normal. Oxygen is what Waters calls a "potential difference maker": it would have made a difference to whether there was a fire or not if it only had differed.

Hart and Honoré ([1959] 1985) expressed the same point without using the language of actual or potential difference maker by stressing that there are contexts of occurrence in which oxygen is a difference maker of a fire, and there are contexts where it is not.[8] It depends on the case—as the case is in the world and as one picks it out—whether oxygen is actually varying. The following was their example:

If a fire breaks out in a laboratory or in a factory, where special precautions are taken to exclude oxygen during part of an experiment or manufacturing process, since the success of this depends on safety from fire, there would be no absurdity at all in saying *such* a case in saying that the presence of oxygen was the cause of the fire. (Hart and Honoré [1959] 1985, 35, emphasis in the original)

Causal Selection and Human Nature

Hart and Honoré's context of occurrence is Mackie's causal field. Since the context also determines which facts are normal, one beats multifactoriality (i.e., that there are always many causes involved) a bit at least: one can leave out everything that is normal and focus on the (in the context of occurrence) actual difference makers—causes that actually differ compared to the other situations in the context of occurrence.

There is a third way how the normality principle can be expressed, used by Hesslow (1988) and closer to the language used in this study so far. The explanatorily relevant cause depends on the choice of what he calls *reference class*, that is, the kind of fire one looks at. Many kinds (though not species) are defined by typical (i.e., statistically normal) characteristics. Thus, the question is: Why was there such a fire [rather than no such fire]"? The terms *there* and *such* point to the assumptions about the respective context of occurrence. These in turn point to the reference class (the type of fire we want to explain) and thus to a specific phenomenon: fire in a lab with no oxygen normally present, or fire in an environment with oxygen normally present.

I now bring the examples together. Explaining an incidence of malaria that occurs in a causal field where people normally are not immune to malaria, one puts nonimmunity in the background. It is not making a difference in such a situation. Rather, it is the mosquito bite that is quoted as cause; (non-)immunity is a potential difference maker, a background condition only. For an incidence of malaria with a reference class that also includes people who are immune to malaria, one puts (non-)immunity in the foreground, since immunity is then a difference maker for whether malaria occurs. The answer to, "What caused malaria in Anton?" would be, "That he had no immunity against it," rather than, "Because a mosquito bit him." Equally, when one explains a fire in an environment that normally contains oxygen, one puts oxygen in the background; when one explains a fire that occurs in a container in a laboratory that normally operates on a vacuum, one puts oxygen in the foreground. Only in the latter case is oxygen a difference maker and thus visible as a cause. Choosing the reference class fixes what is normal and what differs. It thus decides which causes are made visible. What is normal depends on the reference class, which is selected within a context of inquiry, relative to what Van Fraassen (1980) called "why questions." With respect to normality one can thus always ask:

"Normal [only a background condition/potential difference maker] *where and for whom?*"

Making a Phenomenon "Due to Nature" by Choice of Context

The essential point for this study is that the choice of context of occurrence can make something "due to nature." Gannett (1999) quotes lactose intolerance as an example from Hesslow (1984) that is similar to the malaria example and makes explicit how the pragmatic choices of reference class makes a trait "due to nature":

> Lactose intolerance is considered to be a genetic disease in Northern European populations where ingestion of milk products is common and lactase deficiency rare. In African populations, where ingestion of milk products is rare and lactase deficiency common, it is considered to be an environmental disease. … Hence, no trait is "genetic" in any absolute sense, but only relative to a population." (Gannett 1999, 354)

Recall how I have used the lactose (in-)tolerance example in previous chapters: it is a pragmatic decision whether lactose intolerance or lactose tolerance is part of human nature, given that none is truly typical for all contemporary human populations. This chapter adds that if one focuses on kinds of people for whom milk drinking is normal, one will foreground the differences that genetic or epigenetic factors make and consequently regard lactose intolerance and tolerance as due to nature in the explanatory sense, as due to biologically inherited developmental resources. If one focuses on a different group of people, the trait will appear to be due to nutrition.

The context of occurrence refers to something in the world (normal where?), but it also refers to a choice (normal for whom?). Contexts of occurrence are in the world (i.e., the population one chooses for the reference class is real) and determine something somewhere as normal, but it is somebody who chooses this reference class rather than another one as relevant for the explanation. With that choice of reference class, a polymorphic trait such as lactose (in-)tolerance can be made to appear due to the pool of biologically inherited developmental resources.

9.2 Choosing among Actual Difference Makers and the Willingness to Control

So far, all the examples I have used are of the sort already described in chapter 4: different people choose different differences and different reference

Causal Selection and Human Nature 197

classes. To understand causal selection in a narrower sense, one needs examples where people disagree about the cause of something despite the same reference class (i.e., the same population and the same difference assumed). In other words, one has to find an approach that deals with examples that have more than one actual difference maker in a specific context of occurrence and where a choice is made between them. The task is then to explicate how people select between actual difference makers.

For the sake of the argument, I take the following from Collingwood to be an example with more than one actual difference maker:

> Suppose that one medical man can cure a certain disease by administering drugs, and another by "psychological" treatment. For the first the "cause" of the disease will be definable in terms of bio-chemistry; for the second in terms of psychology. (Collingwood 1940, 306)

Collingwood (1940, 306) adds that what the psychologist does and believes "simply records the fact that cases of the disease have been successfully treated by psychological methods." The same can be assumed for the biochemically treating physician. Collingwood treats this case as evidence that "two persons who can treat the same disease in two different ways will make different statements as to its cause" (306). Imagine it as a case with truly the same reference class. One can, for instance, assume that the two doctors have been working for years in the same clinic, always seeing the exact same patients. They thus look at the same reference class and at the same difference—a patient having the disease at time $t1$ and not having it at time $t0$. Still, each doctor picks out a different cause that is involved in the occurrence of the disease, and it is the one he or she was previously able to intervene on in order to cure or prevent the disease.[9] For the psychologist, the biochemical mechanisms are like the hills in the car example. A psychologist cannot and does not care to intervene on biochemical, molecular mechanisms in the same way that a car driver does not care to intervene on a hill when a car stops on it. For the biochemically working physician, by contrast, beliefs and desires are like such hills. The physician cannot and does not care to intervene on beliefs, desires, and the like. The situation is functionally as in the car case: like the driver of the car, she does not care to intervene on beliefs and desires since they appear to her as hills—things that might be causally relevant but are explanatorily irrelevant for her, given the means of intervention (i.e., controllability) available to her.

For areas such as medicine, education, social domains, and technical domains where controlling things is an issue, "practical sciences" as Collingwood called them, he concludes that

causal propositions ... will ... be in essence codifications of the various ways in which the people who construct them can bend nature to their purposes, and of the means by which in each case this can be done. (Collingwood 1940, 307)

Given the clarifications on normality and reference classes, the conclusion is that in cases where the reference class is the same, there might still be a choice to be made among those things that are actual difference makers. Controllability can be involved in that choice, as it is in the example I just used from Collingwood. On top of normality, controllability allows choosing among those causal factors not yet relegated to the background by normality considerations. This is one sense in which Collingwood's approach is more inclusive than Hart and Honoré's normality approach; it also covers cases of causal selection between actual difference makers. But there is another sense, which I will discuss once a serious problem that Collingwood's approach faces is dealt with.

A Problem for Controllability

Collingwood ignored that we sometimes call something a cause (and foreground it), even if we cannot practically intervene on the cause. Ability to control is thus not necessary for foregrounding, it seems.[10] Geneticists, for instance, focused on genes long before they had the means of experimental control on genes. Ability to control is not even sufficient for foregrounding, since there are things that we can control and background nonetheless. There are "nongenetic factors both internal and external to the organism [that] are amenable to experimental manipulation" and that are backgrounded nonetheless, as Gannett (1999, 358) stressed. Oxygen supply in explaining the development of organism is an example. The list of cultural and environmental factors (relevant for the development of an organism) is indefinitely large, and there are indeed many environmental factors that still are (and may always stay) beyond our experimental control, partly because the environment is so incredibly complex, much more complex than the genome or the epigenome. Thus, the representation of environmental causal factors will often be, as Kitcher (2001b, 402) says, characterized by "fragility" rather than controllability. Yet some environmental factors (e.g., water or oxygen supply in plants) have been easy to control from a time when people did not know anything about genes, a time when genes were still pure hypothetically established theoretical entities. These environmental factors were often used to standardize the environment—because they were so easy to control—in order to have a controlled setting for testing the effect of differences in genetic factors. They were controlled in a conservative sense, used as mere standing conditions, fixed and controlled in the background of the "genetic theater," stabilizing what is going on, by creating a normality for the agents on stage. Normality, in other words, is produced by control and can thus be reduced to control.

Willingness to Control
In the light of this, Collingwood's approach needs to be revised toward the willingness to control rather than the actual ability to control. One has to be willing to control for something to select it as explanatorily relevant.

Let's go back to the fire case. What is peculiar about oxygen (in contrast to the kids playing with matches) is not only that it is normally around; it is also the case that we want to make sure that it stays normal and control for it in that sense. We care for it to be present, in sharp contrast to the kids carelessly playing with matches, which is a behavior we aim to prevent. Thus, we put to the foreground those things that we want to control in a forward sense, that is, change in a controlled manner such as the kids' behavior. To the background we put factors we do not care to change or even want to stay as they are (e.g., the presence of oxygen). These are factors which we do not care for or for which we want to control in a conservative sense so that they stay as they are.

Conservative control is making sure that something keeps running as it normally does because it is supposed to run like this. Conservative control prevents change or reproduces the normal state. This is a keep-things-right or set-things-back-to-right kind of control, a conservative kind of control: one tries to prevent something from changing. Yet controlling things can also mean that one aims at changing something. I call this *forward control*. It is important to keep this dual nature of control in mind. If somebody does not care about something or wants to prevent something from changing (conservative control), attention to that something will be hindered and it will be put in the background of causal explanations in the practical contexts at issue here; if people want to forward-control something, they will pay attention to it and include it in causal explanations. Something stays normal because we control for it in the conservative sense (because that is what we want), and if something is not yet normal, we try to forward-control it so that it becomes normal (i.e., the way we want it). Even if we cannot practically change something (control in Collinwood's own terms), we foreground it and call it a cause as long as we are willing to intervene on it, and vice versa. The willingness to forward-control not the ability to control biases us toward some causes as the difference maker of our choice.[11]

The thing that we are prepared to intervene on goes to the foreground, and the thing that we want to stay normal goes to the background. What we are willing to intervene on (or at least are prepared to accept as being intervened on) is foregrounded, whereas what we want to keep normal (with as low as change as possible, because we do not need background control over it or have it already) is backgrounded.

This is then how norms make causes visible: the deviant is the devil (i.e., the cause), since norms (based on social conventions, technical possibilities and interest) define what "we" (usually those in majority or power) are willing to control, that is, willing to set or keep right, and this not only guides us in selecting among the causes; it in turn creates or stabilizes what is normal.

Applied to Prioritizing Biologically Inherited Factors

The revision of Collingwood's control principle helps in understanding why molecular causes (such as genes) rather than cultural or environmental ones are so often singled out in contemporary science and society, even if everyone (when pressed) agrees that genes are ontologically on a par with environmental or culturally transmitted developmental resources, and thus equally causally relevant for producing human traits.

In other words, with Collingwood, one can explain the molecularization of life in science as well as society, despite the fact that it is always nature, culture, and environment interacting when a trait develops in an organism of whatever sort.[12] If one has the means to control the environmental factor, one might still background it; if one does not have the means to control genetic factors, one might still select them as "the" cause for a disease. It all depends on what one cares for and in which sense. Geneticists care for genes since it is their job. By extension, molecular biologists care for molecular causes for the same reasons. Cultural anthropologists have a different job and therefore see different things.

Normative Aspects and Disagreements in Causal Explanations

We ignore, for instance, oxygen as a cause of fire (in the standard contexts of occurrence) because it is something that we care for in life; we want it to remain as is: normal. Because of that, we do not care for it explanatorily. Causal ignoring (backgrounding) or causal selection (foregrounding) is then not just a matter of facts, but also a matter of policy—a matter about what ought to be done (or not done), such as intervening with the children in the fire example to not do again the things they did.

There are normative issues on which people in a society agree. Some become norms: explicit or implicit agreements with normative power to care for or do things in certain ways, such as to drive on this or that side of the road or that children should not play with matches. Norms make (some rather than other) causes visible and explain why many causal explanations seem so self-evident in their partiality ("intuitive," some philosophers would say). It is the agreed-on norms of a society that make it correct to ignore oxygen and explain the fire in the barn by reference to the kids playing with the matches. It is thus not a transcendental higher kind of cognition (or intuition) that makes the fire example so convincing. It is our widely shared social norms and other background assumptions that make it correct to take the kids as the cause and not the fire.

Nevertheless, in every society and in every context, there will also be disagreement on what ought to be done. What some people care for might not be what other people care for. Thus, agreement or disagreement about what people care for (technological possibilities, interests, norms) accounts at least in part for agreement or disagreement about causal explanations and, consequently, about who or what is to be praised or blamed (i.e., held responsible) in case something went wrong (e.g., a disease, accident, illegal, or immoral act occurred). In all cases, the deviant—the disliked anormal—is the devil, the cause responsible for things having gone wrong, whether it is children (rather than oxygen) in daily life or genes (rather than environments) in sciences and society more generally.

To sum up, the willingness to control explains relevant agreement in causal explanations as well as disagreements, by reference to agreement or disagreement on what one appreciates to be changed (foreground) or cares for or accepts to stay normal (background). In that sense, Collingwood's principle of the relativity of causes holds, even though on the basis of willingness to control rather than on the basis of controllability itself.

9.3 How Norms Make Human Nature Visible

In this section, I give, first, a concrete historical example on how social attitudes about what should stay as is and what one is willing to intervene on correlate with shifts in how something is explained. I then extrapolate from the example to a general reflection on how the willingness to control makes sense out of why genes were a focus in the twentieth century. Both examples are meant to look at prioritizing the explanatory human nature over its contrasts, culture or environment.

Shifts in Regularity Regimes of Cancer Prevention

The norms about what should stay as is (and thus be in the background of causal explanations) with respect to regulatory regimes of cancer prevention historically changed in interesting ways, at least according to historians Proctor (1995) and Schwerin (2011).

During the Atomic Age after World War II, awareness about the carcinogenic effects of nuclear radiation, new chemical substances, and air pollutants increased. Well into the 1980s, a dominant political answer (for dealing with the risks that the respective technological progress brought with it) was a regulatory regime that controlled for emissions of radiation and substances by setting limit values. This regulatory regime treated individuals as vulnerable passive objects of irreversible, harmful, mutational effects of carcinogenetic factors penetrating individual bodies. The general regulatory rationale in policy was protecting citizens. What caused cancer? "Industry!" was the dominant reply. Proctor (1995, 171) calls this explanatory regime "body victimology."

The situation slowly changed during the 1980s when, despite new political regulations, three developments came together. First, more and more scientific evidence became available that there are individual differences in cancer-relevant cell repair mechanisms (Schwerin 2011, 143–149). Second, control for environmental and chemical hazards turned out to be very hard: there were too many of them, predictive value of mutagenic tests were unreliable, and, third, "many of the artificial risk factors were so tightly connected with the demands and benefits of modern life that their removal was impracticable from economic and social standpoints," as Schwerin puts it (2011, 150). We got used to and dependent on the environmental causes of the higher risk of cancer.

As a consequence, the vulnerable citizen became an active force to be governed over time. The new regulatory rationale was then "body machismo," as Proctor (1995, 171) calls it. The political imperative became governing the bodies: citizens were asked to boost their repair mechanisms by eating fresh vegetables and buying products from the health industry such as artificially produced micronutrients. If people developed cancer, it was because they did not boost their self-regulation machinery properly. The cause of cancer was now increasingly regarded as residing in individuals who acted in correct or wrong ways. Increasingly in the background of the causal attributions was the environment, containing radiation and substances emitted by a growing industry of energy, food, and pharmaceutical production. That industry moved to the background of causally explaining cancer incident rates not only because of the interests of those profiting from it, but also because it served a consumer culture to which people became used to and dependent on. The majority of the people did not want to change their consumerism.

In the language used by toxicogenetics, as Schwerin reports, the detrimental environmental influences that had once been conceived as "bullets," hitting passive citizens, became conceived as mere stimuli, relegated to the background of political regulation regimes, a mere "biochemical signal triggering a cascade of molecular reactions" in normal human beings (Schwerin 2011, 147). "Normal" humans, with their evolved repair mechanisms, who took care of themselves, would not get cancer. That was the new regulatory slogan. From then on, "the organism was not a ready-made victim of fateful radioactive or genotoxic disturbances; instead, here was an organism which was built to handle those environmental factors" (148). This organism finally became the "entrepreneurial self" of the "new bourgeois lifestyle" of health and sustainability, a "flexible self that works around the clock to compensate for exposure to altered environments … busy handling risk factors" (156), while others engage in business producing the environmental hazards in the background.

What is the cause of cancer? Environmental hazards? Or individual differences in explanatory nature, that is, differences in biologically inherited developmental resources influencing the repair of cell damage? It depends on what one is prepared to politically intervene on: the toxin-emitting industry or individuals who have to face those toxins.

As we all know, some people have to face much more of these toxins, which is how people of color, for instance, end up having higher rates of cancer: a gene for a certain skin color (rather than a cancer-specific gene in this group of people) leads to social discrimination (racism), which increases the probability of ending up in harmful environments, which in turn increases the probability of cancer in the individual who has to live in this environment. That these changed risks can be epigenetically inherited, as described, for instance, by Kuzawa and Sweet (2009), adds to this crooked and fatal social cause of actual biological differences among groups of people. Natural inequalities can be the product of previous social inequalities: biological race the product of social racism.

That those ruling are not willing or not able to change the social inequalities makes the inequalities in social environments invisible: social inequalities have become the hills in the background of many causal attributions. If asked what causes cancer, it is predominantly the differences within individuals that are identified in such a "body machismo" paradigm of molecularized life. Yet the causal structure in the world (in contrast to the causal field represented and the reference class chosen) has not changed: there are environmental factors, social structures, and culture, in addition to biological differences in repair mechanisms (nature). They are difference makers that are ontologically on a par.

The willingness to control thus explains why it is likely that the sciences of contemporary Western society predominantly investigate which difference genes (and other biologically inherited factors) make, because then "they" (those in power, including the scientists) do not have to change anything about the structure of the respective society or the inequalities in access to environmental and cultural resources.

Research on Genetic Causation in General

The willingness to control also explains why there is so much research on genetic causation in general, why there is the molecularization of life. Geneticization of traits (i.e., explaining something as being due to genes) is one form of molecularization, which often involves medicalization, because genes are causal factors that are molecular and internal to bodies.

When medicalization is involved, financial interests are too. As Gannett (1999, 359) wrote, "The United States government has been motivated to fund the Human Genome Project for the sake of the health not only of the American people but of its developing biotechnology industry." "Blaming genes," she states,

> draws society's attention away from unhealthy environments and weakens its commitment to address factors such as poverty, cigarette smoking (and tobacco advertisements), exposure to pollutants, and racism [racism as illustrated above, not race!], which all contribute to these diseases. (Gannett 2008, 455)

Kitcher made a similar point as Gannett, even though it is the risks of the family members of the affluent that define the context in his example:

> Those of us who live comfortably worry less that our sons and daughters will suffer from neglect or abuse—we, after all, intend to provide them with safe and nurturing environments. But we are not immune to disaster. The genes may strike, and if they do, all our effort will be in vain. Accordingly, we are very interested in one kind of cause of the diminution in quality of people's lives. Other kinds of causes, environmental factors that wreak havoc with lives, are not (perceived as) *our* problem. (Kitcher 1996, 311; cf. 2001b, 131–132; emphasis added)

Those who define the problems, define the causes. This is the logic of control: there are social reasons for backgrounding environment (inequality as not important to change, or even important to keep), and there might be further reasons for foregrounding genes (things some people want to have forward control over).

What somebody is prepared to control, as well as what is easy or convenient to intervene on, depends on who it is and what that person cares for. Geneticists, the pharmaceutical industry, and those well off care for handles they can or hope to be able to intervene on, because they get refutation from it, or money, or both, or because this is the only thing they cannot control yet, or because they are driven by some theoretical or personal belief in the moral or social importance of this or that. Gannett (1999) mentions the example of the scientist Jerome Lejeune, who not only discovered the genetic cause of Down syndrome (caused by an extra chromosome 21), but went on to try to find other causes by studying the mechanisms how the condition comes about given the extra chromosome. His opposition to abortion motivated his search for nongenetic causal factors of Down syndrome, as Gannett reports. Whether or not a baby will be born, it seems, was not what he was prepared to intervene on. His goal was consequently "to find some other 'handle' by means of which to intervene in the treatment or prevention of the symptoms associated with Down's syndrome" (367).

9.4 How Norms Make Human Nature Real

So far, I have analyzed how norms make causes visible and thus how norms or normative considerations of individuals decide which partial explanations are given, with a few hints as to how racism is causing biological differences. This section shows in more detail how the latter works, that is, how norms make causes real and, in particular, how norms make human nature in the explanatory as well as the descriptive sense.

Here is how our explanations in part create who we are. In step 1, considerations about what should stay as normal as it is (conservative control) or change (forward control) lead to causal selection. In step 2, this leads to partial explanations. In step 3, since explanations also carry claims about responsibility, people may react to these explanations in a way that stabilizes or changes their respective behavior. In step 4, this stabilizes or changes the distribution of traits (the pattern of similarity or difference), and thereby the normality at issue. The context of occurrence (and, with it, the reference class of the next event of the same kind) is thereby stabilized or changed. The pattern of similarity and difference of the metapopulation—the descriptive human nature—is made thereby, not by construction alone (i.e., purely epistemically) but by an explanatory looping effect, that is, by literally changing the kinds of people existing in reaction to the explanations prioritized. Since the change in descriptive nature, step 5, can change the normative considerations as well, the process has an iterative, looping form (back to step 1).

The term *looping effect* has become somewhat virulent since Hacking (1986, 1995, 2007a) used it for describing how people react to classifications, for example, to being classified as homosexual. People react to a classification, he claimed, by moving in or out of the classification, changing the kind thereby. I use the term *explanatory looping effect* for an analogous effect of explanations and in a manner that allows for rather indirect, intergenerational looping effects. I will illustrate the explanatory looping effect with reference to three examples that I have already discussed: the toy explanation of why there was a fire in the barn, the explanation of Down syndrome, and the explanation of racial differences in cancer susceptibility.

Three Cases of Explanatory Looping Effects

The explanation that the children (rather than the oxygen) caused our case of fire will likely (if it is effective) lead to disciplining those children. It thus influences how they will act in the future. If the behavior of the children in the population changes, the explanation has led to a situation where the children move out of the explanation of such kinds of fires, so that they cannot be held responsible anymore. They are set to what is expected from them, what is normal. Explanations for fires will then become less likely to point to kids playing with matches; after all, there are fewer kids doing that as a result of the disciplining. If so, then the willingness to control them has led, first, to causal selection, which has led, second, to a partial explanation, which has led, third, to different behavior by the children, which has led, fourth, to stabilization or change of the traits exhibited in the population at issue. This can, fifth, influence (stabilize or change) the norms that guide people in their explanation of practices: once the disciplined kids are adults, they might repeat what happened to them, disciplining their own kids in turn.

If the case were one of self-discipline, the looping effect would be more direct, but I do not regard directness as decisive for a looping effect. What is decisive is that reflexivity is involved: it must be humans influencing the pattern of similarities and differences in humans by their partial explanations to regard a consequence of a partial explanation as a looping effect. Not any consequence of explanations will do. Imagine that the distribution of malaria-transmitting mosquitos changes as a result of some interventions, given the partial explanation of malaria pointing to mosquitos. Such an effect of the explanation on the distribution of mosquitos in the world would be a causal process that changes the prevalence of a kind, but it would not be a looping effect in the narrow sense since no reflexivity is involved. After all, the change in distribution does not stem from mosquitos giving explanations and reflectively changing their behavior accordingly.[13]

The looping effect can also be intergenerational, as the Down syndrome example illustrates. If one foregrounds the chromosomal defect and intervenes accordingly, this explanatory partiality can change the patterns of similarity and difference in the world. If molecular factors (such as the chromosomal number) are foregrounded, then abortion of fetuses with the chromosomal defect becomes more likely. This was precisely why Lejeune tried to find other causes: to have other kinds of interventions available. Selective abortion, or any genetic selection (call it eugenics, call it enhancement), changes the pattern of similarity and difference in the world. It thus changes the very constitution of the reference class entering in subsequent explanations of the respective population. The population will be normalized away from differences with respect to that one trait, creating homogeneity of the kind at issue (analogous to disciplining the population of kids in the fire case). Those opposed to enhancements are already painting the specter of such eugenic homogenization. A new stable typicality, a new aspect of the descriptive nature, is thereby made real. Certainly that process can never be complete, as scientists stressed long ago against eugenics' ambitions to "clean up" human nature. After all, genes are much harder to discipline than kids. Note that I am not arguing for or against enhancement (that is a difficult ethical issue). All I want to show is that norms influence which explanations are given, and this in turn influences what kinds of people exist, which includes what kinds of developmental resources are frequent and widespread and which are not. This explanatory looping effect, even in its intergenerational form, applies not only to cases of enhancement.

In the case of the different cancer explanations, for instance, the backgrounding of environmental factors (in the sense of conservative control) will further stabilize the embodiment of racial inequalities in the respective society; it undoes human nature, so to speak. It strengthens patterns of difference (rather than homogeneity) that will show up at the descriptive level (which patterns of similarity and differences one can observe among humans) as well as the explanatory level (which patterns of similarity and difference one can find with respect to developmental resources involved in the production of the phenotypic traits that one can observe).[14]

The Normativity Involved and Human Nature on the Move

All of these are cases that involve contested normative considerations. By negotiating them continuously, we will continuously change and continue to become human (in the classificatory, descriptive, and explanatory senses). We continuously become human in an open-ended, reflexive manner that depends not just on adaptation to some environment but on how humans think humans should be. Human nature is a reflective project of humans, always on the move.

Normative considerations determine how things should be and thus what needs to stay as is or how things should become. Normative considerations control in that dual sense, and they influence causal selection, guiding us in our partial way of dealing with causes, be it in a scientific context or a more general social context. They determine our socially preferred ways of intervening, which determines causal selection. Since sciences work in a social context, it is sometimes the social context that shines through scientists' willingness to control, as in the case of the scientist Lejeune and his opposition to abortion. People go for the difference makers that they are willing to intervene on and are guided by considerations of normality in doing so. Because it is for most humans normal not to ask whether oxygen might be omitted when we want to explain the occurrence of a fire, it is for many unfortunately still normal not to ask whether inequality or other social phenomena might be omitted when we want to explain diseases, IQ, antisocial behavior, and the like. In the background are all those things one does not want to change (even unconsciously), such as inequality. By putting them in the background, ignorance about what would happen if we were to change them is produced. If we do not intervene on the environment (e.g., the nutritional one in the norm of reaction for body height), then we do not see the difference it makes. The choice of possible worlds that we imagine in our scientific and social affairs is in that sense decisive for the attributions of responsibility we make and also decisive for what I called explanatory looping effects and thus for what kinds of people will exist.

Normative considerations (e.g., how humans should be) make causes not only visible but also real. Over the long run, causal selection is literally involved in making human nature by changing the distribution of the developmental resources and traits of humans at issue.

Finally, it is not the case that sometimes normative considerations are involved in our causal explanations and sometimes not. They are always part because any partial causal explanation involves choices about what one is interested in and thus also choices about what one is prepared to change or not.

To conclude, human nature is a concept with a normative dimension (how humans should be) that influences not only what is considered normal (as described as part of the dehumanization challenge) but also what is considered causally relevant and, via explanatory looping effects, what kinds of people will exist in the future. Human nature, in the descriptive as well as in the explanatory sense, at each point in time contains the causal factors humans were not prepared to change. This is how, in part, we constantly and literally create human nature.

Summary of Part II

The resulting post-essentialist, pluralist, and interactive account has three natures surviving: a classificatory, a descriptive, and an explanatory nature.

Since the reference of the term *human* can be to humankind or humanity, the three natures of the scientific image are superimposed onto each other, resulting in two times three natures in the world. There is humankind and humanity and their respective classificatory nature, each giving rise to a descriptive and explanatory nature of the respective group, as depicted in table 9.1.

The relationships between these concepts of human nature are complex: the descriptive nature needs the classificatory nature, either with respect to humankind or humanity, in order to have a clear reference class; the explanatory nature explains the descriptive nature if "important" is interpreted (as in contexts of evolution, heredity and development) as "stable" even though it is—if production of the traits is at issue—even in such

Table 9.1
Two times three post-essentialist natures.

Kind of nature and related question	What it refers to	Where it is used
(1) Classificatory nature: Who are we and who counts? (a) Humankind (b) Humanity	(a) Human species (genealogical nexus) (b) Moral community	(a) Evolutionary biology, phylogenetics, but also social sciences, humanities, philosophy (b) Social sciences, humanities, philosophy
(2) Descriptive nature: How are we?	Human life form (generalizations about humans)	Descriptive fields of life sciences (e.g., anatomy), social sciences, humanities
(3) Explanatory nature: Why are we the way we are?	Biologically inherited developmental resources	Explanatory fields of the life sciences—in particular developmental biology and developmental psychology

contexts not the only kind of cause involved, as the developmentalist challenge reminds us.

The classificatory nature of humankind is determining the genealogical channel of biological inheritance, the channel, in which developmental resources partaking in the explanatory nature are traveling from generation to generation. Therefore, the classificatory nature of humankind is also explanatorily relevant for the explanatory nature of the humankind and indirectly also for the descriptive nature of the humankind. The nature-nurture contrast can easily be applied to the explanatory nature and, though more difficult, to the descriptive nature, but it is possible by focusing on difference making (rather than production) and allowing for reconstitution of phenomena via abstraction.

An animal-human boundary is relevant for the classificatory nature but not necessarily for the descriptive or explanatory nature: traits do not have to be species specific to belong to the descriptive or explanatory nature of humans. Species-specific traits are epistemic identifiers (e.g., useful for the field guide of the hypothetical Marsian scientists coming to Earth to study humans) but otherwise not prior. Nonetheless, the resulting post-essentialist pluralism allows for adding that one is specifically interested in traits that allow making a human-animal distinction by looking at typical traits that are exclusively typical.

An alleged normative nature is not part of this scientific image except as foundation for the explanatory looping effects. If the explanatory looping effects occur, humans over time become what the majority of humans think they should be. Since this means that norms, which are part of culture, make natures, it is, together with the interactionist picture developed in chapter 4, why the account developed here can count as truly interactive. In part III, I build on the results of part II and suggest that the term *normative nature* should be used simply as the label for what people agree on with respect to how humans should be. But since that claim and the picture developed in part II are not dependent on each other, I discuss it separately.

The resulting post-essentialist, pluralist, and interactive account of a new scientific image of human nature has the following characteristics. It takes into account that the term *human* historically has always been used in a perspectival manner: the group that speaks takes itself as the paradigmatic exemplar of being human. Furthermore, even within one such group (e.g., the WEIRD), there is a biological and a social way of fixing the group identity. Which group is chosen depends on what is of interest, which—in the case of scientific interests—depends not only on the society in which the science is produced but also on disciplinary affiliation (e.g., biological or not), personal interest, and curiosity, and thus on the questions asked.

The account adequately situates the importance of genealogy, roughly as Hull suggested, but with a pluralist bent. After all, classification is not all that humans or even scientists care for and the importance of genealogy for delineating a group, biological or not, is a contingent fact. We could do without it. We could cease to regard our identity genealogically.

The account gives typicality of traits and their stable covariation (i.e., their clustering) a constructive place by acknowledging that only it, if connected with a suitable account of what makes the cluster stable, can fulfill the contrastive descriptive role that essences traditionally played: not too narrow (making human nature disappear by definition) but also not too broad (making everything humans are, think, and do a part of human nature by definition).

The account manages to integrate the common distinction between nature and culture by taking the causal explanatory role of a "nature" seriously, but without falling back on any explanatory essentialism, with its intrinsicality bias, and without ignoring that culture is ontologically on a par with nature in developmental, epigenetic, and evolutionary timescales. It thus does not lose sight of the interactionist consensus in its developmental, intergenerational, and evolutionary dimension. Last, but far from least, the account has no need for any normalizing capacity talk, which would only boost dehumanization practices.

Part III discusses further questions not dealt with so far: it addresses what is left of the alleged normative nature of humans and whether it would be better to eliminate the term *human nature*, given what's left of the concept.

III Normativity, Essential Contestedness, and the Quest for Elimination

This part discusses what follows from part II for how science, values, and society interact. Part II already mentioned some of the connections. The three kinds of natures established in part II are real but relevant for what it means to be human only depending on the following pragmatic, evaluative, and social issues. That our sciences care for genealogy (in determining species' boundaries and membership in them) is contingent and very likely socially influenced in a way that might well change in the near future. Which among typical properties are regarded as important also depends on scientific and social preferences. Whether our explanations ignore nature or culture is dependent on our scientific interests and social values since it is normative considerations on how humans should be that guide causal selection. This can lead to explanatory looping effects, the making of parts of the explanatory and descriptive human nature.

There are two further important normative issues, also resulting from the pluralism defended, that I have not yet addressed. First, what happened to the traditional idea of a normative essence, referring to a bundle of properties humans should have? What is the ontological status of that kind of nature? Does it exist? Which role and function does it play? Can one say more than that it feeds, on the one hand, into dehumanization and, on the other hand, into looping effects? In other words, is there any humanism left if one follows the parsimonious pluralism advocated here, assuming only what needs to be assumed to classify, describe, and explain humans and their properties in a manner as objective as possible? Finally, should scientists eliminate any reference to the term *human nature*, given that it has become so much more ambiguous as part of the pluralist frame? Wouldn't it be better to speak about the three natures in separate terms to prevent confusion and the smuggling in of essentialist baggage? If one can divide things up the way suggested in part II, why not use different labels for the two times three natures? Isn't clarity—as an epistemic value—forcing one to opt for elimination of any human nature talk given that it became so ambiguous as part of the post-essentialist frame? Epistemic values such as clarity will be juxtaposed with social values, such as preventing dehumanization, as part of discussing that final question in chapter 11.

10 Humanism and Normativity

I take humanism to involve normative issues about moral standing, justice, care, and human rights. What follows for humanism, given the post-essentialist, pluralist, and interactive account developed in part II? The first aim of this chapter is to show that it does not pose any special problems with respect to these normative issues. The second aim is to show that the alleged normative human nature of the traditional picture can be reduced to a subset of the properties that make up the descriptive nature. The third aim is to show, based on the ideas of Gallie (1956), that human nature in the normative sense is essentially contested; it is a concept with respect to which there cannot be a consensus, not because of different perspectives involved but because of the dialectic character of what is denoted with the concept. The contestedness is in the concept itself and thus essential to it. It is part of the concept to be contested.

In section 10.1, I return to the comparison between humankind and humanity that I introduced in chapter 1. I defend that despite the pluralism advocated here, there is only one moral group of humans, which is important for issues of moral standing and human rights. Showing this will include some notes about speciesism. In section 10.2, I discuss in which sense the normative nature can be specified and how it can be reduced to the descriptive nature despite its normativity. In section 10.3, I dwell on the dialectic aspect of the normative nature, which is connected to the explanatory looping effects from chapter 9. This then leads to the claim that human nature in the normative sense is an essentially contested concept.

10.1 Two Sufficient Entry Conditions for Moral Standing

Humankind and Humanity

I assumed at the beginning of this study that there is humankind and there is humanity. Kelsen had already written:

> Society and nature, if conceived of as two different systems of elements, are the results of two different methods of thinking and are only as such two different objects. The same elements, connected with each other according to the principle of causality, constitute nature; connected with each other according to another, namely, a normative, principle, they constitute society. (Kelsen 1943, vii)

The elements are human individuals. Their form of connection, a genealogical nexus in the case of humankind and a social nexus in the case of humanity, defines who belongs. The criteria of membership are in both cases contingent, and with respect to the classificatory nature of humanity, this contingency is absolutely key since the social group is clearly also a moral group: it is a group of people who have moral standing and are subject to justice, care, and rights. "Contingency" in this context means that it is the choice of those already in the social group to determine the entry conditions for it.

The social group (Kelsen calls it "society") does not and should not rely on any biological membership criteria. Any biological membership criterion would decide membership for the social group for the wrong reasons, which would be analogous to nationalism and racism; it would be "speciesism" as the standard dictum in moral philosophy goes. I take speciesism to imply using membership in a species simultaneously (i.e., monistically) as a membership criterion for humanity. It would be a continuation of nationalism and racism since it would give or deny moral standing and rights for morally irrelevant reasons, that is, reasons derived from biological facts such as genealogy.[1] Because these biological facts are in and of themselves morally irrelevant, I kept normativity out of the new scientific image developed in part II.

Although it follows from the last point about speciesism that the two group categories are distinct, it does not follow that they are completely different or unconnected. On the contrary, they are closely connected by overlap, as illustrated in chapter 1. Some, if not most, individuals who are members in the biological group are also members in the social group. But the overlap is not one by entailment. Neither group is a subset of the other.

Humanism and Normativity

After all, as discussed in detail in chapters 3 and 5, there are members of humankind that are clearly not members of humanity. In turn, there can be members of humanity who are not members of *H. sapiens*. Take your preferred humanoid robot or the superchimp of recent philosophical imagination. Humankind is thus not a subset of humanity, nor is humanity a subset of humankind, even though there is a huge actual overlap between the two groups. That overlap has been stable in the past, and I reckon that it will persist for a while—despite all the superchimps, hybrids, and robots populating certain areas of scholarship and culture and despite humans who think they can raise their children as if they were squirrels.

The Quest for One Human Moral Group

If one distinguishes between humankind and humanity in the way I suggest, there is still the question of whether one ends up with two different moral groups of humans. It seems that we need more than overlap, some kind of strict unity, for political or moral purposes and for the vision of humanism that characterized the Enlightenment and gave rise to such ideas as distributive justice, global ethics, and human rights.

If the concept of being human in the social sense is regarded as decoupled from the biological sense of being *H. sapiens*, only showing overlap, then the gates are open to regard some *H. sapiens* as not (enough) human in the social sense: children and others who lack the respective properties for membership in the social group (e.g., because of severe physical or mental disabilities, or because the individual is at the margins of life, that is, the beginning or end) can thus be dehumanized—regarded as less human in the social sense. Doing so (e.g., as part of the reflections of Singer 1975, 2009, and McMahan 2002) has received considerable critique, especially from the perspective of disability studies (e.g., by Kittay 2008 and Kittay and Carlson 2010).[2]

Without going into the complexities of the debate about moral standing, I claim that an important step toward a solution is a double-entry condition for moral standing. "Marginal" humans can be regarded as fully human in a different sense, namely, a biological sense. Being human in the biological sense can be treated as equally sufficient for moral standing as having the respective traits that are regarded as essential for the social group, such as rationality or feeling of pain.

In the following, I can only outline the reasons that speak for such a solution. A full treatment of this vexing issue is beyond the scope of this study. The aim here is not to defend the double-entry solution, but to show that in and of itself, the pluralism of this study does not conflict with the traditional humanism assumed in moral and political philosophy: that we should not exclude any human being from having equal moral status.

The Double-Entry Solution to the Problem of Moral Standing

The double-entry solution states that there are two sufficient conditions for having full moral standing: you are in if you descend from other *H. sapiens* (first sufficient condition) or if you are like humans, which requires that you are similar enough to be able to interact in morally and politically adequate ways with other humans (second sufficient condition).

It is a fair question to ask why there is a broad consensus that the so-called "marginals" should have full moral standing. I have no ultimate answer, except that it is a choice that is historically contingent in the sense outlined in chapter 5, a choice that takes genealogy as morally and politically important without being speciesist. We care for where we come from genealogically, and parent-offspring relations have a special moral status. Special responsibilities result from it, at the individual and group levels, codified in many laws, across the globe. They are the background of genealogy's importance in understanding what it means to be human.

The question why we cannot just exclude humanoids by genealogy, that is, why the social group is not simply decided by genealogy as necessary and sufficient, has already been answered: that would be speciesism. Genealogy has some moral importance, but it does have limits. It does not have the power to kick someone out of the moral group. I do not know of any reason to exclude a humanoid that is morally, emotionally, and cognitively like humans from getting human rights (e.g., the right for bodily integrity, which includes protection from being killed). To think otherwise would indeed be a moral speciesism that is as unjustifiable as nationalism.

Note that I am committed to a view that takes moral relations as decisive for inclusion in humanity. After all, there can be two distinct justifications as to why certain humanoids should get moral standing: one giving a relational property priority, the other giving similarity priority. A similarity account would take the moral and mental capabilities of the humanoids in and of themselves as decisive. A relational account (similar to the genealogical account of humankind) includes a humanoid person on the basis of whether it is likely that the person will be able and willing to stand in her or his respective social relations to other humans—whether the person will conform to the rules about social interaction, irrespective of what makes it possible for that individual to do so.[3]

The double-entry solution states that if similarity and social interaction point in different directions, then the decisive issue should still be specific ways of social interaction. Whatever an individual is made of, wherever it comes from, whichever social relations it had in the past, as long as it actually relates to other humans in socially adequate ways, it is a human in the social sense. Similarity is diagnostic only: a hint at the social way of being human, of being humane. It would be evidence of the possibility of adequate social interaction but not the criterion that decides membership in the moral group. Similarity in and of itself (i.e., that a humanoid looks like us, is made out of flesh like us) is not what counts, since what is the moral relevance of such a similarity in and of itself? The moral relevance of certain similarities can only stem from the fact that they (the ability to reason and to speak, for example) enable respective morally relevant social interactions. But if that ability is ontologically realized differently, why should that matter? A moral philosophy that grounds moral standing in similarity itself is as much subject to a comparison to racism as grounding moral standing in genealogy is subject to the charge of speciesism as analogous to racism. It is just a different kind of racism at issue, often now called "ableism" and related to property-based dehumanization.

To sum up, the solution I suggest for moral standing is a double-entry solution: membership in each, humankind or humanity, is sufficient for being a human in the moral sense, sufficient for moral standing. For the pragmatic function of deciding about moral standing, one can combine the two overlapping populations of humans, humankind and humanity, in an adequate way. This solution is accomplished by having neither kind of classificatory human nature as necessary for being a human but each as sufficient. As a result, there is only one moral group of humans, even if there are two different concepts of being human: a biological and a social one.

Certainly this is far from solving all the problems that moral and political philosophy struggles with in connection with moral standing, justice, and similar other issues. The unitary account should result in all humans having the same human rights, but it does not decide the question of whether all thus included humans have the exact same rights and duties in all respects. Parents do have special duties regarding their children, and children usually do not yet have the right to vote, and so on . Nonetheless, they are all protected by human rights, and even special rights of children. Issues about these differences in special rights, which include issues about the so-called ethics of killing, cannot be answered in this study.[4] It would lead too much astray into the depths of moral and political philosophy.

10.2 The Ethical Importance of the Descriptive Nature

A second issue from moral and political philosophy needs to be taken into account. After all, membership is not all we care for when humanism is morally defended. We also care for ways of dealing with people, of treating them. Given this, the question arises, Which are the traits that are traditionally called "naturally good" in the sense that a human being, a member of the moral circle, should have it for its flourishing? How do we arrive at the list of traits that belong to such a normative nature? I will discuss different positions, which I call normative essentialist and internal normativist. I will criticize the first and show that my account is compatible with the latter without deciding between versions of the latter.

Normative Essentialists

Normative essentialists (e.g., Foot 2001 or Thompson 2008) believe that there is a way to derive the respective normativity of human nature directly from insights about a species. According to them, a species norm stems from a species' normative essence. Yet the descriptive nature that survives the three challenges from part I will not itself provide such a normative essence since it is just descriptive. Where is the normativity of the normative essentialist then coming from? I see two options.

Normative essentialists can mount it on a version of an intrinsic essentialism of shared capacities, yet they will then fall with that rotten fundament (as discussed in part II). Alternatively, they can refer to a concept of normal functioning. That solution has been addressed by Amundson (2000) and, more recently, Odenbaugh (2017). The problem is that the concept of normal functioning is in itself highly contested and ambiguous. As mentioned earlier, evolution regards everything as normal that survives. In addition, the concept of normal functioning furthers dehumanization, for instance, by not distinguishing the level of functioning (e.g., in terms of communication) and the mode of functioning (e.g., spoken rather than sign language): disabled people often show an equal level of functioning (e.g., in communication), even if their mode of functioning is different. I take the critical arguments of Amundson and Odenbaugh with respect to the concept of normal functioning as convincing.

The one option left will not please normative essentialists: one can refer to social deliberations that decide which of the typical and stable traits are important. I will defend in the following that it is these deliberations that ground the normativity that comes with a humanism of sorts, an internalist humanism.

Internalist Humanism
The normative properties of a good human life can be defended as derived from an internally normative concept of human nature. The foundation for the normative nature then consists of social deliberations that result in a choice about what is important among the traits typical and stable. Nussbaum (1992, 1995, 2000, 2006) has been most outspoken in defending such an internalist humanism. She explicitly opposes reference to an "external" normative essentialism that derives the normativity from outside ethics, from the order of the world, that is, matters of fact that scientists study. Her argument is that if we agree that there should be global ethics, justice, and human rights, then there needs to be a concept of being human that refers to a list of needs and capabilities that we regard intrinsically (i.e., intrinsic to normative reflections) as normative and simultaneously as part of a descriptive human nature. This normative human nature includes, for instance, a need for food, drink, and shelter; the ability to have a language; the ability to reason; and the ability to affiliate with other humans. The last two are, according to Nussbaum, "architectonic, holding the whole enterprise [the human life form] together and making it human," in a similar way as in Aristotle, for whom reason and sociality played a similar central role. These needs and capabilities need to be guaranteed for every member of *H. sapiens*, so that they are able to fully participate in the human life form, the descriptive human nature. Yet which traits belong to the normative human nature also depends on our ethical visions of how we want to be. Humans select among the facts of the matter what counts for a good human life.

Nussbaum (1992, 227) distances herself from the concept of personhood, which she treats as less "determinate" than the concept of being human, even though both concepts are intrinsically normative for her. Interpreting her account in detail would require much more space than available here. It is unclear, for instance, how Nussbaum relates the human life form to humankind. On the one hand, she seems to assume the membership problem to be solved since for her, "every offspring of the human parents" (1992, 228) counts as having the basic capabilities essential for the human life form, until evidence of severe cognitive disability or the like proves otherwise. On the other hand, she seems to defend a monistic essentialist account since she regards the descriptive nature as simultaneously classificatory. Her account still has the "determinate account of the human being, human functioning, and human flourishing" (descriptive nature) as conveying certain "central defining features" (Nussbaum 1992, 205). I will leave it open how to best characterize her account with respect to these issues. I instead compare her approach with a more liberal position and then show that my account is compatible with both versions of an internalist normativism toward the human: Nussbaum's account and the more liberal position.

Veil-of-Ignorance Humanism

The liberal tradition, deriving from Rawls (1971), is equally internalist but has a different pattern of argumentation attached to it: a veil-of-ignorance pattern. It takes the concept of personhood as the foundation of ethical reflections and usually defends the neutrality of moral theory regarding biological-anthropological facts (those about the biological species). Nonetheless, under a veil of ignorance, a list of primary goods is derived that every person should have access to. The way the list of morally important parts of the human life form is arrived at is thus internal (as for Nussbaum) but still different from Nussbaum's way, since the latter also involves asking real people. In terms of the ethical importance of the descriptive human nature, however, the accounts amount to the same: they internally select a list of ethically important parts of the human life form, but they construct the list based on knowledge (of real or imagined people) about what that life form is as a matter of fact. The normative nature is thus a subset of the descriptive nature and in that sense can be reduced to it. A life worth living for humans is a human life.

Compatibility with the Internalist and Veil-of-Ignorance Humanism

Irrespective of differences in moral theory between Nussbaum and the liberal account, I take my approach to be compatible with both, as long as both accounts allow membership in humankind or humanity as sufficient for belonging to the moral circle. I take them both to implicitly assume (or at least have to assume, for consistency) that there are facts about a descriptive human nature. These facts enter the ethical reflection, be these Nussbaum style or veil-of-ignorance style.[5]

As Antony (2000) argued in reaction to Nussbaum, one cannot be too internalist with respect to human nature since then the concept becomes too culture dependent, that is, not global enough to reach a consensus about the interventions in justice, care, and issues of human rights. For a global ethics about distributive justice and human rights, one needs objectivity about the human life form. The only objectivity that is available is, however, the one pertaining to the scientific facts about the descriptive and explanatory nature of humankind: genealogy, typicality, stability, biological reproduction, developmental resources, and the like. Even those deciding about the list behind their veil of ignorance cannot afford an ignorance about these facts, despite the neutrality of their perspective.

Nonetheless, and here I agree more with Rawls than Nussbaum, the implicit or explicit scientific image is, first, not in and of itself moral and, second, it allows going beyond it and imagining how else humans could be. It allows for looping effects by also deciding how humans should be. In the veil-of-ignorance picture, the explanatory looping effects that make vivid how humans make their nature can be imagined more easily than in Nussbaum's picture.

But in both cases, Rawls's and Nussbaum's, the respective resulting lists are at each moment in time a subset of the descriptive human nature established in part II, even if that subset can be a result of the ethical reflection and is also one of becoming human rather than merely being human. Certainly, looping effects did not create that humans need food or shelter to survive, but they are involved in some of the capabilities Nussbaum arrives at in her list of capabilities of human life, such as the capability to affiliate with members of the same species. Given the hostility many refugees fleeing to Europe have experienced, it seems that currently, there are plenty of humans in Europe who have not yet developed that basic capability of the human life form. Any call for civility (e.g., to treat refugees adequately) is a call for a looping effect, a call for staying or becoming human. Given the history of dehumanization, there seems to be a need for the kind of reflexivity involved in such looping effects. Moral capabilities need to be stabilized; humanness is not guaranteed by an internalist heaven, a metaphysical heaven, or scientific facts. We need to recreate it constantly. In the spirit of B. Williams (2006), the claim is that the foundation for ethics is not provided by religion or philosophy; it is recreated by humans every day. In other words, it evolves.

The Paradox of Ethically Relevant Human Universals
Antony (2000) has stressed that even if one grants what I have stated so far, there is still an ethical paradox to be solved: either properties of the human life are ethically irrelevant or not typical. Take the need for food as an example. That humans need food is ethically irrelevant since people want food that is not disgusting and what is disgusting varies from culture to culture. For some people, certain kinds of meats are disgusting, for others it is other kinds of meat, or even meat in general that is disgusting. There is certainly an abstract generalization that still holds, namely that humans need some food, as they have some language. But that humans need some food is ethically not what is relevant, that is, it is not sufficient to make ethical decisions.[6]

The same context dependency exists for other central issues in global ethics—for example, the kind of medical treatment and the kind of gender roles. What is adequate and needed with respect to these needs will not be typical in a straightforward sense across all human populations.

In reply to such context dependency, we can, however, abstract away from the differences (e.g. in the need for food) by creating a new typical property: that people want nondisgusting food. The solution is similar to the one I discussed with respect to language in chapter 7. Humans typically speak a language, irrespective of which precisely. Certainly, which food is disgusting still varies from culture to culture, but it is a typical trait that people want some nondisgusting food. Nussbaum (1992, 224) argued that what is typical exists on a "high level of generality" that is always in need of concrete "specific and historically rich cultural realization."

Thus, before giving up a global ethics, we should make that leap toward the abstract and ask people in each case which food is disgusting for them and act accordingly. There is nothing illusionary in that kind of descriptive abstractness and nothing colonial. It provides one with typicality and moral relevance simultaneously. One just needs to hit the right level of abstraction. This is a practical solution to the paradox and in any case a better solution than giving up humanism just because it involves abstraction. Thus, from a moral point of view, it is not enough to know some typical trait, for example, that humans need food. One sometimes has to upgrade the general knowledge so that it becomes morally relevant.

The theoretical paradox, though, holds in a comparative form, despite the practical solution: the more you concretize the typical trait, the less typical it gets, and the more you keep it typical, the less ethical relevance it contains. It fails to match what people care about.

To sum up, which traits a human being should have is derived from a choice regarding which of the typical and stable properties are morally or politically important. The normativity stems from choices since it comes in via the selection of traits that are already part of the descriptive human nature or are envisioned to become so. It is nonetheless still a matter of fact, at each point in time, what is on the list of typical traits from which one can choose the important ones.

Reliable knowledge about the traits belonging to the descriptive human nature is hard to get, and it is even harder to get knowledge about the developmental resources needed to sustain it. To make the sciences studying humans reliable and unbiased is as important as with every other kind of knowledge. Eternal vigilance is the only responsible reply. Yet to condemn all scientific endeavors that study humans in general as unreliable in principle has no justification. After all, it depends on the details of whether specific approaches in the human sciences are methodologically reliable. These details are of utmost importance, but they have nothing to do with the concept of human nature.

10.3 A Dialectic, Essentially Contested Concept of Human Nature

I conclude that human nature in the normative sense is a set of objectively existing (or possible) traits that are important in a specific sense: in the sense that a group of people discursively agrees on them to count as decisive for a flourishing life. They are traits a human being should have, given a moral community negotiating and agreeing on that vision.

Human Nature as a Dialectic Project

Since there are looping effects involved, I take it that this normative nature is what Cassirer meant when he wrote the following:

> Only by way of dialogical or dialectic thought can we approach the knowledge of human nature. ... It is not therefore like an empirical object; it must be understood as the outgrowth of a social act. ... Man is declared to be that creature who is constantly in search of himself—a creature who in every moment of his existence must examine and scrutinize the conditions of his existence. In this scrutiny, in this critical attitude toward human life, consists the real value of human life. (Cassirer 1944, 6)

Rorty (1999, 52) similarly wrote, "Humanity is an open-ended notion, ... the word human names a fuzzy but promising project rather than an essence." Human nature in the normative sense is a project, always on the move, and it is contested. Similarly, Fukuyama (2002, 128) wrote, "Due to the intimate connection that exists between human nature, values and politics, it is perhaps not surprising that the very concept of human nature has been extraordinarily controversial over the past couple of centuries." The reason it is controversial is its essential contestedness.

Human Nature as an Essentially Contested Concept

There will be—at each moment in time—more than one suggestion for the normative selection from the descriptive human nature. There can then always be a plurality of suggestions available for the normative human nature, and given there is such a plurality, there will be contestation of the right meaning (the right subset of the descriptive nature and the right projections).

It is, however, not necessarily the case that concepts are contested because they lack one clear interpretation. It can just be the other way around: concepts may lack a clear meaning because they are contested, and that may be essentially, or necessarily so. The political scientist and philosopher W. B. Gallie (1956) suggested that there are such concepts that are tenaciously ambiguous for a reason that has to do with the concept itself rather than with differences in knowledge or psychological or ideological motives in using the concept. He called such concepts essentially contested ones: "concepts the proper use of which *inevitably* involves endless disputes about their proper uses on the part of their users" (Gallie 1956, 169, emphasis added). I will describe that special sort of concept and argue that the normative concept of human nature is such an essentially contested concept.

Gallie (1956) wanted to elucidate cases where disagreements about concepts persist, despite the matters of fact being clearly on the table. For instance, despite intensive and long-lasting debates about the meaning of concepts such as democracy, justice, or truth, no consensus has been reached about their meaning. Also, what a gene is, a case related to the topic at issue in this book, is contested, despite large agreement on the matters of fact involved. But not every contested concept is essentially contested. With Gallie, one has criteria to further differentiate between contested concepts that are essentially contested and those that are only contingently contested.

One such criterion sticks out: the criterion of having an exemplar. It is the criterion that puts polysemic, ambiguous concepts aside (i.e., concepts that connect to only one word but have different meanings depending on interest, perspective, and the like). For an essentially contested concept (contrasted with a merely ambiguous one), there must be at least one exemplar that everybody contesting the concept accepts as falling under the concept. "Democracy," according to Gallie, has such exemplars. Yet there simply is no such exemplar, for instance, as I would claim, for the concept of a "gene." There might be a core of the concept in an abstract sense, as Waters (2000) suggested, but that does not entail that there is a concrete string of DNA that is considered a gene in all the different versions of gene concepts that are now distinguished. Does the normative nature of humans have clear exemplars? I think so: there are, after all, core capabilities that are morally not contested (e.g., the need for nondisgusting food).

This leads to three further important criteria that Gallie mentions: even concepts that have a clear exemplar (and are thus not just polysemic) can be internally complex in having different parts, different fillings that can be weighted differently. Internal complexity is important for a concept to be essentially contested. It is the source of a third criterion: initial contestedness. An essentially contested concept needs to be contested from the very start of its use. I do not see any problem with this criterion. Fourth, the concept as well as the phenomenon meant with it must develop in an open manner, that is, in a manner that "admits of considerable modification in the light of changing circumstances; and such modification cannot be prescribed or predicted in advance" (Gallie 1956, 172). That condition is likely to be fulfilled more easily by things that involve looping effects, so it is safe to assume that it applies to the concept of human nature. Fifth, there need to be some acknowledgment of the contestedness by the involved contesting parties. It is hard to find out whether this condition is fulfilled in examples such as democracy, truth, or human nature since the parties contesting the concept are unbound; ultimately it is the complete group of all humans, humanity and humankind taken together, past, present, and future. But I do not think that this condition is so decisive; in any case, it is less decisive than the final two.

According to Gallie, an essentially contested concept needs to be, sixth, appraisive, referring to an achievement. He uses, for instance, championship in addition to democracy as an example. Since I specified the normative nature to be the nature that is contestable, it should be clear that those who contest it in the way discussed here use the concept of a normative human nature in an appraisive manner; it is only that each appraises something else.

Most important is Gallie's last condition, the condition of productivity of contest. For a concept to be essentially contested, the disagreement must be productive: by continuing to discuss the content of the normative nature, the negotiating community needs to come closer to what is generally meant by the concept, independent of or maybe even because of the concept's inherent complexity, openness, and initial contestedness. Human nature, in the dialectic normative sense discussed here, does exhibit this kind of productivity. Combining the language of Gallie, Cassirer, and Rorty, human nature is a productive dialectic project. "What it means to be human" in the normative sense is that by endlessly contesting the normative subset of the descriptive nature, we endlessly further develop as humans.

Eliminating a Contested Concept Misses the Point

Often none of the stakeholders is prepared to give a contested concept away. Yet there are usually also those who choose the ultima ratio of the eraser minded: they deny that the concept makes sense and—in the name of clarity or parsimony—suggest getting rid of it. There is no justice; there is no truth; there is no human nature. Certainly if a concept is eliminated, nobody can claim it for the respective interests and nobody can do any harm with it, but giving up an essentially contested concept is missing the point of its contestedness: having disagreement about a contested concept is productive for achieving something of value that is connected with the concept. That holds for the concept of human nature as it holds for pluralism of attitudes with respect to democracy. The assumption behind the analytic category of essentially contested concepts is that the endless debates about the meaning of truth, democracy, or human nature make it more likely that something of value connected with what is contested is achieved.

This is the magic in essentially contest concepts that eraser-minded philosophers fail to see. Disagreements, ambiguity, and internal complexity of concepts are thus not enemies of philosophy. If taken straight ahead, they might help in understanding what is going on, in general and especially with respect to these intricate, intimate, and deeply rooted concepts such as truth, democracy, justice, and human nature.

Nonetheless, even given the essential contestedness of the normative human nature, I will argue in the final chapter that it might be adequate to eliminate the term *human nature*.

11 Should We Eliminate the Language of Human Nature?

This chapter asks whether scientists should still speak in the name of human nature given the diversity of ways in which this can be done. After all, alternative terms are available: *human condition, human life form, developmental resources, classificatory criteria,* and others.

With the help of five claims, I will defend an eliminativist perspective with respect to the term *human nature*. The first claim is that sciences themselves would not lose much for the understanding of the matters of fact involved if they stayed away from calling the genealogical nexus or the descriptive or explanatory regularities "human nature." Yet, second, there is no way out of dehumanization since getting rid of essentialism and the language of human nature would not be sufficient to prevent dehumanization. Nonetheless, third, post-essentialism and eliminating human nature talk are steps in the right direction in order to minimize dehumanization. In addition, fourth, the question about elimination versus revision of the terminology used is actually a matter of values (rather than facts). Seeing this allows to apply, fifth, a precautionary principle that directs us toward elimination. All things considered, I will claim, it is better to prevent human nature talk as much as possible. The price of such a linguistic elimination is low, and the risk of damage (in terms of dehumanization) will be too high if human nature talk is not eliminated.

In section 11.1, I introduce the issue about elimination and revision as mostly about words (rather than concepts). I also discuss why I treat elimination as a regulative ideal (rather than an achievable goal). I then show in section 11.2 that the term *human nature* has become not only ambiguous but also redundant. Using one term for three (or four) different matters of fact has no justification if describing matters of fact is at issue. I defend that dehumanization is minimized even though not prevented by going post-essentialist and by getting rid of the term *human nature*. On that basis, I introduce in section 11.3 three trade-offs between science and society and suggest a precautionary principle for application.

11.1 Elimination versus Revision

Elimination versus Revision as Two Rhetorical Strategies about Terms

As a leftover from essentialist thinking, the term *human nature* is in danger of becoming metaphorical only, as often happens with terminological leftovers. For instance, the language of having this or that "in the blood" is still used, but is not taken literally anymore. Yet scientists such as Galton literally believed in the claim that developmental resources are biologically inherited via the blood. For them, it was a fact, not a metaphor. To take statements about something being in the blood to be literally true today would simply mean to anachronistically fall back to an outdated nineteenth-century theory. Is, thus, using the label "human nature," for the genealogical nexus between humans, or for the life form of humans, or for the bundle of biologically inherited developmental resources, equally anachronistic? Part II showed that there are states of affairs that one can call a human nature. But should we call these things that way? It all depends on the consequences of a language of "human nature." Is it beneficial? Is it harmful? And according to which values?

Elimination as Regulative Ideal

Elimination of a term such as *human nature* (or *race*, for comparison) is treated in this chapter as a regulative ideal rather than an achievable goal since success in linguistic elimination is unlikely to be successful. I thus distinguish a normative elimination question from a descriptive elimination question, which addresses whether linguistic elimination is likely to be successful if tried.

Success in eliminating the language of human nature has two dimensions: success in eliminating the term and success in thereby eliminating the potential misuse of a historically and politically laden concept. It would be instructive to conduct and compare a couple of historical case studies on how calls for linguistic elimination versus revision of similarly contested terms (e.g., *race*) have fared. I cannot offer such a case study here and do not know of a decisive answer in the literature for the expression *race* or *human nature*. There are some good starts in the right direction with respect to *race*, which suggest that the call for elimination was unsuccessful, in both dimensions.[1] I thus assume that any call for eliminating the term *human nature* will equally fail to be completely successful in both dimensions (eliminating the term and thereby preventing negative effects of the concept).

Yet complete success is not the only thing that counts. Minimizing the negative effects of a concept by pruning our ontology and language is a worthy goal. This is why I want to discuss the issue in a normative manner independent of any historical observation or prediction about the success of such an endeavor. At issue is thus whether we should at least try to eliminate the term *human nature* in order to minimize dehumanization.

11.2 Redundancy, Neutrality, and Risk of Dehumanization

Redundancy of Human Nature Talk

Given the pluralism of the post-essentialist account developed in this book, it depends on the respective field which successor notion is scientifically adequate. There will never be agreement on whether the classificatory, the descriptive, or the explanatory concept of human nature is best since each option caters to different but equally important disciplinary needs. In addition, there is a normative nature, prevalent in political and moral philosophy. So they should all be regarded as equally valid successor notions.

Moreover, the language of human nature is dispensable for all of the post-essentialist successor notions. One could refer to all the three post-essentialist human natures (descriptive, explanatory, and classificatory) in the sciences and also to the normative nature in philosophy without ever using the term *nature*. Given that it is not necessary to use the expression *human nature* to talk about humans in any of the senses developed in part II and chapter 10, the price of eliminating the language of human nature can be taken to be not very high. It thus holds that sciences and philosophy would not lose much for the understanding of the matters of fact and normative choices involved if they stopped calling the descriptive or explanatory regularities or the genealogical nexus a human nature.

Neutralizing Is Not Possible

Chapter 2 on the dehumanization challenge established that what we think about human nature influences how we treat other beings we consider as not or less human. I will use the distinction between property-based and relational dehumanization from chapter 2 to claim that getting rid of essentialism is not sufficient for getting rid of dehumanization. There is no way to make sure that as part of the post-essentialist frame, human nature talk can become politically neutral and thus free of any dehumanizing potential. Neutralizing (i.e., sanitizing) the language of human nature is not possible.

To review the results from chapter 2, people get dehumanized in the sense that properties deemed to be part of humanness are differentially attributed to members of the human species. Standard cases of that kind of dehumanization include women, ethnic groups, people with deviant sexual or social behavior, and poor people. There are other cases of dehumanization that are rather heritage based, predominantly connected with racist thinking. What is differentially attributed in the first place is heritage (rather than properties, which are only indicative of heritage). The exclusion of Amerindians or Africans as belonging to a non-Adamic heritage is a well-known historical example.

Both forms of dehumanization have a post-essentialist human nature notion corresponding that can further dehumanization. The purely descriptive concept of human nature can be involved in property-based dehumanization, and the genealogical concept can be involved in relational dehumanization. The risk of dehumanization remains even if stripped off essentialist baggage. Even if one restricts the use to post-essentialist concepts and does so without any human nature talk, the problem of dehumanization will persist. The problem lies deeper, much deeper. Let me develop this claim for both kinds of dehumanization.

Even if scientists use only a descriptive concept of humanness, dehumanization cannot be quarantined. A minimal descriptive concept merely describes how humans at a certain historical time are on average or by majority worldwide. (It is more minimal than what is defended in part II since it leaves out the time dimension and thus also stability as typicality over time.) Although human nature then only refers to contingent generalizations about humans at a certain time, without assuming an essence (without any classificatory, explanatory, or normative force attached to the generalized properties), it will still have dehumanizing potential since individuals who do not conform to the generalizations can still be dehumanized. Any case of ableism or contemporary eugenics is a case in point for such nonessentialist dehumanization.[2] As long as those in power, or the majority, rule (descriptively and socially, in science or society), those not in power, or not in the majority, can face some discrimination in terms of being regarded as less human. Thus, dehumanization can still take place on the basis of a minimal, nonessentialist concept of a descriptive nature.

A genealogical eliminativist approach would admit that there is (as a matter of fact) a genealogical nexus between humans that allows classifying them as one humankind. We can refer to the genealogical nexus without ever talking about that nexus being a shared classificatory human "nature." As with humanness, as a matter of clarity it is actually advisable to use the more specific language of genealogical nexus than the more generic talk about a human nature. But even if one therefore completely abstains from traditional essentialist assumptions as well as from the language of a "nature," there is still a danger of dehumanization in connection with the genealogical nexus. After all, the genealogical nexus can, independent of essentialist thinking and the term *human nature*, be used for exclusion from a moral circle. Neither speciesism nor racism depends on a traditional essentialism. Whenever genealogy is used to include some as members of one and the same kind (e.g., by Darwin to fight racism, as described by Desmond and Moore 2009), it can simultaneously be used to exclude others. Historically, the circle drawn by the genealogical human prejudice grew, but the kind of inclusion or exclusion principle remained the same.[3] If genealogy is used for inclusion or exclusion, only those connected in more or less direct genealogical ways are included in the respective social circle, whether the circle is about human rights, property rights, or what have you. Thus, whenever genealogy is used as an inclusion principle, it can exclude some individuals as not human.

Today that dehumanizing potential matters only theoretically, not ethically. Despite the much discussed claims of the technical as well as philosophical enhancement industry, all individuals whose status as humans is actually contested are long dead.[4] Therefore, the only real expulsion boundary is the vague boundary of our species' beginning. Yet genealogy allows for grading within the one humankind since the genealogical relationship can be more or less direct. Contemporary dehumanization of black people (e.g., as discussed in chapter 2 and mentioned above) can be explained by (and would be predicted) on the sole basis of a belief in a different or at least more distant heritage of a particular black person to a particular white person, compared to the distance between that person to another white person. A belief in an individual's genealogical distance to another individual might even (but does not have to) involve a belief in genealogical races (i.e., separate groups united only indirectly deep in history by a common human ancestor). Dehumanization exists without racism, but it often uses racism.

Whether a genealogically motivated dehumanization combines with a genealogically motivated racism or not, the more or less close heritage can be sufficient for dehumanization to happen. As long as genealogy can be used to create distance, it can be used in a dehumanizing manner. "The more closely related, the more human" would be the logic of that variant of post-essentialist relational dehumanization. It is a matter of empirical and historical study whether such a form of dehumanization is still widespread and whether it explains historical cases, but I assume that it can exist and that it is likely to be the basis of some forms of contemporary racism, given that genealogy has historically been used for dehumanization of ethnic groups with a specific geographic origin (e.g., as in the case of the Amerindians as non-Adamic in heritage discussed in chapter 2). I conclude that as long as genealogy is used for delineating kinds and allows for grading, the genealogical nexus can be used to regard some people as less human, independent of essentialism and independent of the term *human nature*.

The politics of human nature might be resolved by eliminating the term *human nature*, but the politics of being human will stay simply because there is an independence of dehumanization from any essentialist baggage and the language used to talk about humans. Even if one gets rid of all possible essentialist baggage attached to human nature talk, and even if one gets rid of all human nature talk whatsoever, there is no way to make sure that the concept of being or becoming human gets rid of dehumanization. Stripping off essentialism and the language inherited from it won't suffice for that.

In contrast to Smith (2011, 2014), I thus argue that there is no essentialism necessary for dehumanization to happen.[5] This is so because in social life, the concept of being human can be specified in a functional manner only, as stated in chapter 2. The vernacular concept of being human can be filled with all kinds of different content, and these contents can have essentialist connotations. This is why the problem is deeper than the problems stemming from essentialism. Given the functionalist account defended in chapter 2, it even looks as if no pruning of our language and ontology will ever completely get rid of dehumanization because any content can function in a dehumanizing way.

I conclude that completely neutralizing post-essentialist concepts of human nature as well as the language we use is not possible. There is no way out of dehumanization since getting rid of essentialism or the language of human nature is not sufficient to prevent dehumanization. The problem of dehumanization is deeper.

Post-Essentialism and Eliminating Human Nature Talk as Steps in the Right Direction

I nonetheless support the claim that post-essentialism and eliminating human nature talk are steps in the right direction in order to minimize dehumanization. Essentialism discounts variation much more than the post-essentialist successor concepts and is therefore much more dehumanizing. A post-essentialist picture is thus more equalizing since every member of the human species is regarded as equally normal and equally abnormal.

If post-essentialist successor concepts, already less dehumanizing than the essentialist concepts, would in addition drop the language of human nature, it is likely that this would prevent dehumanization even more because of the essentialist baggage of that terminology. If only the outdated concepts, but not the terminology, are eliminated, the probability of misuse by equivocation is high. Imagine that psychologists show that there is this or that regularity (e.g., heterosexuality as stably typical of the species) and thus calls it part of human nature (rather than just a statistical generalization about humans).[6] It is easier to derive that the respective behavior is "normal" and that humans should behave the way if the language of human nature is used. This is so simply because of the history of the term *human nature*, which carries with it a significant essentialist and therefore normalizing and normative baggage. This is why I think even post-essentialist concepts can more easily be misused in a dehumanizing way if they use the language of human nature.

Talk about being human is still dehumanizing, even though talk about human nature (because of the essentialist baggage attached to that language) is likely to be much more dehumanizing. Therefore, post-essentialism and eliminating human nature talk are steps in the right direction to prevent dehumanization as much as possible.

11.3 Elimination versus Revision as a Matter of Values

Three Trade-Offs between Science and Society

Aren't the claims made so far providing a solution for the elimination question? After all, the price of eliminating human nature talk for representing matters of fact is low, and eliminating essentialism and human nature talk together are more effective in minimizing dehumanization than eliminating essentialism only even though it is not completely preventing it either.

The problem is that there are some trade-offs involved that make the issue a bit more complex. First of all, with respect to the scientific price for elimination there is a trade-off regarding epistemic issues that are only indirectly connected with representing matters of fact. With respect to minimizing dehumanization, there is a similar trade-off that concerns positive consequences of using the term in social affairs. These positive consequences need to be taken into account too. The core of the problem is, as I shall argue, that the elimination question is a matter of values, not just facts. Both stances, an eliminative stance and a revisionary stance, are value-laden attitudes.

• *Trading off epistemic values*. First, clarity is an important epistemic value, be it in philosophy or science or elsewhere. If clarity is our guide, then post-essentialists should, even irrespective of dehumanization, eliminate any reference to the term *human nature*. One should rather talk about what one exactly means: a genealogical nexus, a set of explanatory factors that is of special importance in explaining why people are the way they are, or a specific set of reliable generalizations about humans, or the respective normatively established subset of these generalizations. Yet terminological continuity or ambiguity might be of value too, and even for purely epistemic reasons. Keeping old terms (while changing their meaning) creates epistemic continuity and thus unity between those bits of the outdated framework that are still valuable. It helps integrating the old knowledge into a new (e.g., post-essentialist) frame. We would all like that for the sake of itself, right? After all, unity is, like clarity, an epistemic value that is rarely challenged as being in and of itself valuable. Given this second epistemic value, unity, there should either be a clear replacement candidate for what traditionally was meant by the term *human nature* so that the branches of research with still valuable knowledge that was connected with that term in the past can be rescued too. Or, as the more pluralistically minded post-essentialist might prefer, the different and still valuable branches of knowledge, past and present, that can be connected with an ambiguous term *human nature*, should be connected simultaneously, using ambiguity of a term for fruitful translation of research results across a diversity of fields. The term *human nature* would then be like a hub that allows connecting different research fields, past and present. Yet, others might reply, the price is high. All of the clarity that could be gained by eliminating the term or at least fixing the reference is lost. Lots of talking past each other might result. But as others will counter, clarity is not the only epistemic value we cherish. Is there a way to end this back-and-forth, a way to trade off the two epistemic values of clarity and unity? I think there is not: there cannot be an in principle solution for this trade-off, and in any case, it is not a question of mere matters of fact, but a matter of also trading off epistemic values.

• *Trading off social consequences*. Second, the decision between elimination or revision becomes even more difficult once the boundary between science and society is traversed. Politically the concept of human nature has had not solely negative effects. Dehumanization is just one side of the politics of human nature. The concept of human nature also had an important emancipatory equality-establishing function. I pointed to it when I mentioned the history of the idea of one humankind. Thus, Nussbaum (1995), Antony (1998, 2000), and Silvers (1998) are correct in their claim that in order to criticize certain ways of treating humans (so that they are all treated equal and as humans), some content is needed, that is, some descriptive account of what it means to be able to live a human life form is necessary for ethics; otherwise, ethical critique would be impossible. I take Mikkola (2016) to defend a position along these lines, as well. In order to decide about elimination versus revision, one would then have to weigh the effects of equality-establishing against the effects of dehumanization. Disagreement will loom large not only with respect to how many positive and negative social effects have actually occurred in the past, but also with respect to whether the equality-establishing function of the concept and term is still so important that our society (and consequently science as part of society) should stick to the term *human nature* (rather than using alternative terms such as *humanity* or *human condition*). The term *human nature* might well be a Wittgensteinian ladder: a ladder that we needed to arrive where we are (in our dialectic project) but that we can now throw away.

• *Trading off overall scientific usefulness and overall social usefulness.* Third, scientific consequences have to be balanced against social consequences. The problem then is that even if we could agree (somehow) on the overall scientific and social usefulness of the language of human nature by solving the first and the second balancing problem, there is no way to directly trade off overall scientific usefulness against overall social usefulness.

These three balancing problems have to be taken into account when the question is raised as to whether a revisionary or an eliminative stance toward the term *human nature* is adequate, given what's left of human nature as a concept (i.e., given the post-essentialist, pluralist, and interactive account developed in this book).

Most philosophers seem to share the assumption that matters of fact decide whether we should revise or eliminate the term *human nature*. Yet matters of fact alone cannot decide that issue. There is no ultimate fundament in the world for deciding between elimination and revision of the language we are using since both the eliminative stance and the revisionary stance are value-laden attitudes. Since science uses a language that is often value laden, a baggage that accumulates over time, science is in and of itself a normative endeavor. There is thus a politics of human nature in that sense: it is a political issue how scientists talk about humans.

A Precautionary Principle Applied

That answering the elimination question involves trade-offs of consequences and values shows that a consensus on whether one should eliminate the traditional expression *human nature* will be unlikely—in principle at least. Yet in practice, a precautionary principle can be applied.

I think, given that dehumanization is hard to eliminate, it is safe to assume that there is a risk of misuse of the expression *human nature*. The consequences of the misuse would presumably be high (e.g., when homosexuals have to face imprisonment, still the case in some countries in this world). In addition, it could affect half of humanity in cases where women are dehumanized. Thus, in practice, the risk of misuse is high. In addition, we can use a different language (e.g., the language of humanity) for the equality-establishing function if we do not want to trade in that positive side. Thus, we can fulfill equality-establishing goals without ever using the term *human nature*. This amounts to a practical solution of the second in-principle balancing problem.

Finally, there is agreement that the scientific price of elimination is low (for matters of fact, as established in section 11.2). If it is (and I think it is), and if it is also the case that different results from past and present science can be connected without using the term *human nature*, then this amounts to a practical solution of the second in-principle balancing problems.

Given this final observation about the risk of misuse and the price of elimination, a precautionary principle should guide us toward elimination. If the social risks are high and the price of elimination across science and society is low, then precaution should guide us toward elimination as a regulative ideal (solving the third balancing problem). We should try to prevent the terminology as much as possible and, if that is not possible, disambiguate human nature talk whenever possible.

Linguistic elimination will certainly not be a solution to everything. It will neither guarantee that the essentialist concept is eliminated nor prevent dehumanization completely. It would, however, be an important step in the right direction of keeping the politics of being human in check. After all, humans can adapt, and that includes that they can change their way of thinking as well as their language.

It might sound a bit ironic that a book on human nature ends with the recommendation to stop using the term *human nature*. Yet without further ado, I close with the Wittgensteinian move I mentioned above: sometimes we need to climb a ladder in order to be able to throw it away.

Summary of Part III

Part III showed that there are three normative issues that ask for eternal vigilance: who counts, how a descriptive concept connects with normative issues about how humans should live, and which language is used.

There is a descriptive form of life for humankind as well as for humanity. Both humankind and humanity are important for moral standing. The descriptive form of life of the resulting unified moral group relates to properties that those who are in that group regard as important for that life form—traits that humans should have. Via looping effects, the respective normative decisions change the very pattern of similarities and differences that make up that life form (and derivatively, the explanatory resources related to it). What's left of human nature is a dialectically changing project of humans becoming human. Since there is often no global agreement on how humans should be, what it means to be human turns out to be essentially contested. Yet by continuously contesting how humans should be, we become more human.

The term *human nature* should be prevented as much as possible (as I have in this summary). The reasons are that in a post-essentialist world, the term *human nature* has a multiplicity of concepts attached to it, as well as a set of pragmatic functions, some mainly important in science (e.g., creating clarity or unity), some in society (e.g., furthering dehumanization or creating equality). The pull of these pragmatic functions can point in opposite directions, toward eliminating or revising the language of human nature. Yet given that the price of elimination is low and the risk of negative effects of that language is high, precaution shifts the balance toward linguistic elimination. We should stop using the term *human nature* whenever possible.

Notes

Preface

1. Leslie White (1958, presidential address to the anthropology section of the American Association for the Advancement of Science), quoted by Sahlins (1976, 105).

2. In this book, I mostly use terms like *we*, *us*, or *our* and also *West* without quotation marks, except for cases where special care is necessary. Nonetheless, it should be clear that these terms are indexical and thus perspectival (i.e., their meaning is dependent on who speaks).

1 Introduction: What's at Issue

1. Neither the *Stanford Encyclopedia* nor the *Routledge Encyclopedia* contains an article on "nature." Presumably, this is because the concept is too fundamental, too entrenched in too many different fields. For a take on it, see a classic, Mill (1874), or, more contemporary, Collingwood (1945), C. S. Lewis (1960), or R. Williams ([1976] 2011). A connotation that is completely ignored in this study is the contrast between nature (*physis*) and art (*techne*). See Böhme (2002) for a discussion of it.

2. It is the latter that is often capitalized as "Nature." Whenever I am using the term *things* in this study, I aim to stay neutral with respect to metaphysical issues about things, substances, and processes. In particular, I take the account defended here to be consistent with any process metaphysics of human nature, as, for instance, defended by Dupré (forthcoming).

3. See Lloyd (1991) for the invention of nature in Greek antiquity.

4. Note that this is different from asking whether individual organisms have essences (i.e., whether there are individual essences without which the individual would not be that individual but a different one). Discussions about personal identity are about individual essences; discussions about human nature are about kind essences.

5. East/West is yet another divide that some (e.g., Sahlins 2008) consider important with respect to the concept of human nature.

6. See Klein (2016) for the shifting taxonomic use of the term *human* and the latest news on the classification of humans. Klein uses *human* for *Hominini*, which is a name for a tribe. His article also shows that the term *human* shifted its reference so that the term stays exclusive. Traditionally, only humans were placed in the family Hominidae; thus, the term *human* was used for the whole family. Given new evidence about the close similarities between humans and other great apes, the terminology and classification changed. The term *human* moved two steps down: from family, via subfamily, to a tribe. That way, the terminological boundary between humans and great apes stayed: there are humans and there are great apes. All great apes and humans are now regarded as members of the family Hominidae (the former humans only), with two subfamilies, one called Homininae (including chimpanzees, gorillas, and humans), the other called Ponginae (orangutans only). The subfamily Homininae is further divided into tribes, one of which is the human tribe (Hominini) and the other the Gorillini (for gorillas and chimpanzees). See Proctor (2003) on how morality influences where we draw the human line.

7. Kronfeldner, Roughley, and Toepfer (2014) already used the distinction among classificatory, descriptive, and explanatory epistemic role to arrive at a first overview of the varieties of options for a post-essentialist account. Roughley (2011) already pointed to different epistemic roles. Samuels (2012) had a similar list of epistemic roles, although he does not distinguish between epistemic roles and pragmatic functions the way I do. In his list are "organizing role" (being a pragmatic function in my case), "descriptive function," "causal-explanatory function," "taxonomic function" (corresponding what I call classificatory), and "invariances." Machery (2016a) gave a similar lineup but called the latter "limiting function." In my approach, the latter relates to fixity and is part of the specific kind of explanation that human nature provides and thus not a separate role.

2 The Dehumanization Challenge

1. Parts of this chapter relate to major sections of Kronfeldner (2017). Permission for reuse was granted by Elsevier Inc. 2017 (© Elsevier Inc. 2017).

2. Kelman (1973) introduced the term *dehumanization* for the context of explaining mass atrocities. Similar terms like *subhumanization, infrahumanization,* and *dyshumanization* are still used, and there is no overall agreement on usage yet. For review of (and a specific position on) such terminological issues, see Smith (2014). The usage in this study is not in conflict with any of these usages in the current literature, but orders things in a particular way.

3. The feminist Hubbard (1990) used the phrases "politics of women's biology" and "political nature of human nature" to stress that women's biology has a political

dimension and that it is politically constituted. Following her lead, I take the term *politics of human nature* to refer to the "political nature of the concept of human nature" (a title of one of her papers). When we talk about human nature, we are already talking politics in the sense of conceptualizing our social and thus political interactions and the boundaries we draw between people: we conceptualize with it how we interact and should interact (e.g., what is a morally or politically adequate behavior) and who counts for such considerations, for example, who is in (humanizing) and who is out (dehumanizing). Not surprisingly, feminist scholars have been strong in addressing dehumanization; see, for instance, Jaggar (1983), Nussbaum (1995), Antony (1998, 2000), Haste (2000), and Mikkola (2011, 2016).

4. Lerner and Hofmann (2011) noted a similar coincidence (i.e., tension): normality is not connected to typicality in the case of veterinary medicine, whereas it is so in human medicine.

5. Using a generic linguistic form in describing other animals does not change that. Even if generics (sentences that do not use a quantifier like *some* or *all*) are used for other species, they are not interpreted in a similar normatively laden manner in the case of other species. The normativity that is applied to other animals concerns appropriate treatment of animals. In particular, it is about treating animals as the animals they are. So it assumes a nature of the respective animals, and this nature has normative consequences for the behavior of humans with respect to animals, but that does not normatively structure the respective group of animals into a social hierarchy that has individual animals as more or less realizing its nature.

6. See, for instance, Finnis (2011).

7. See, for instance, Foot (2001), Thompson (2008), and Hursthouse (2012). There are certainly differences in the theories of these authors, but these differences are irrelevant for the aims of this study.

8. Bayertz (2009) gives a good overview. Compare Birnbacher (2009) or Reydon (2014). For a direct critique along the lines of the Darwinian challenge, see Odenbaugh (2017).

9. Daston (forthcoming) provides a history of dehumanization with kind essentialism in mind that focuses specifically on the idea of the unnatural. To dehumanize individuals in the sense of imposing a moral hierarchical order on those not fully realizing the nature of the kind by using the category of the unnatural is one version of dehumanization that follows the same logic of demarcation and exclusion described here. I take Daston to agree on that point, despite all the finely grained differences between different forms of dehumanization and important differences even within the history of regarding individuals as being unnatural.

10. As with all conceptual definitions, this involves a terminological choice that I regard as justified by two assumptions: the assumption that concepts that become too broad will be close to empty and thus useless, and the assumption that this is

how the overall use of the term in science and scholarship can best be understood and reconstructed.

11. Smith (2011, 26–29) distinguishes dehumanization from objectifications (being regarded as object rather than subject) and deindividualization (being regarded as an exemplar for a stereotype rather than an individual) and sets a very narrow focus on cases of dehumanization that lead to aggressive behavior against those being dehumanized. I take deindividualization as a precondition for dehumanization and objectification as a mechanistic form of dehumanization. See Haslam (2006) on mechanistic dehumanization. This solution allows pointing at similarities and differences between the different forms simultaneously. Smith's narrower meaning would be too narrow since on its basis, one would too easily overlook the systematic connections between objectifications, deindividualization, and his narrow animalistic dehumanization.

12. I am using the translation from Platt, which is part of the Oxford edition (ed. by Ross and Smith; see Aristotle 1952).

13. Lloyd (1991) stresses that the Greeks already had a universal concept of humankind. The barbarians were also regarded by Aristotle as members of the same biological species, even though they were, as well as slaves and women, regarded as less human.

14. At about the same time, LeBon (1881, 155–159) also speaks about the "infériorité intellectuelle" of women as "trop évidente" and that they "représentent les formes les plus inférieures de l'évolution humaine et sont beaucoup [sic] plus près des enfants et des sauvages que de l'homme adulte civilizé" (157). The passage from LeBon is often quoted in feminist literature, often after Gould's (1978, 365) translation.

15. For a more detailed philosophical discussion of dehumanization of women, see Mikkola (2011, 2016). For an empirical study, see Eyssel and Hegel (2012). See Kourany (2016) for a philosophical take on the never-ending discussions as to whether there is or can be sufficient scientific evidence for women being less intelligent than men.

16. The quotation from Montesinos and further references regarding that history can be found, for instance, in Smith (2011, 77ff.), also with respect to the famous month-long debate between Sepúlveda (an Aristotelian scholar defending that indigenous people are natural slaves) and Bartolomé de Las Casas (defending their equal human status). Homunculi are often portrayed as "little men" created in the laboratory of the alchemist (as discussed in Paracelsus), but they were also discussed as able to grow up, as Paracelsus does in his *De generatione rerum naturalium*. Certainly they were then regarded as monsters. Smith (2011, 78–80) reports that they sometimes were regarded as beings with a human body but no soul and that

Amerindians were portrayed as such homunculi. For a book-length account of the "Atlantic encounters in the age of Columbus," see, for instance, Abulafia (2008).

17. Polygenism (as a belief in separate human species) was certainly only one route to racism and slavery. Thus, all of the above holds even given that racism in general or slavery in particular was often justified without any reference to those enslaved as not being a member of the same species.

18. Although the results have been criticized by Francis (2015), I regard them as solid enough to support the claim that there is implicit dehumanization of African Americans along the human-ape axis. Francis claims (by using his own statistical measure) that the results are "too good to be true" in the sense of not sufficiently limiting type I errors (false positives). Francis's critique of Goff et al. 2008 is a methodological critique about "the choices in experimental design, analysis, theorizing, and reporting." Yet he admits (2015, 7) that Goff et al. (2008), and the other two studies on the same topic that he criticized, are not violating any "standard methods in experimental psychology." Given this and given the metalevel disagreement about Francis's measures (see the debate, following Francis 2013 in the *Journal of Mathematical Psychology*), together with the abundant evidence about dehumanization from history, anthropology, and social psychology in general, I think that Goff et al. (2008) can still be taken as yet another bit of evidence (albeit not in and of itself conclusive) to support the claim that there still is the kind of dehumanization they study.

19. That it does still happen is also reported by Hund, Mills, and Sebastiani (2016). Comparisons between apes and humans are actually quite old, going back to antiquity, as Corbey (2005) also shows.

20. Also: why we dehumanize others might vary. Haslam (2006, 255) mentions "relief of moral emotion, self-exoneration, enabling or post hoc justification of violence, epistemic certainty in the face of non-normative behavior, provision of sense of superiority, enforcement of social dominion." The differences between these motivations are important and a matter of contemporary research (see Haslam and Loughnan 2014); for the context of this book, however, this layer of analysis has to be and can be ignored.

21. A note on psychological essentialism might be necessary on that point. The functional account presented here is not in conflict with a cognitive one. When Gelman (2003) claims that essentialist thinking in children is the result of everyday thinking (e.g., distinguishing appearance from reality, causal reasoning, labeling), then this is compatible with the functional account. Ignoring the functional account, however, leaves out an important dimension that needs to be stressed vis-à-vis contemporary studies of psychological folk essentialism.

22. Often, in particular in feminist or poststructuralist discourse, dehumanization is understood as hierarchical. Haste (2000, 177) writes, "'Otherness' entails

boundaries, exclusions and inclusions policed by categories and rules. ... Otherness is not reciprocal." That might well be true for some cases, but it is evidently not true for the cases of reciprocity mentioned. The construction of dehumanizing otherness can be reciprocal.

3   The Darwinian Challenge

1. Parts of this chapter relate to parts of Kronfeldner, Roughley, and Toepfer (2014), albeit in a different and much more detailed form. Thanks to my coauthors for allowing me to reuse the material here. (© Kronfeldner, Roughley, and Toepfer 2014).

2. The *via* in the last sentence is important since historians, such as R. J. Richards (1987), Amundson (2005), Winsor (2006), and McOuat (2009), have shown that it is a philosophers' myth that it was Darwin (or even Darwin and Galton) who initiated population thinking. Population thinking has a much deeper and complicated history. In accordance with that, I am referring to the Darwinian challenge not with reference to Darwin and Galton as the great men of history who made population thinking possible. They are rather taken as being involved—albeit as centers of intellectual gravity—in a conceptual evolution. The historical approach I am using in this is inspired by Richards's (1987, 559–592) evolutionary model of history. Furthermore, I am not assuming that those who assumed types prior to Darwin (e.g., Linneaus), in order to establish a natural system of biological entities, did so because of an outdated Platonic or Aristotelian essentialism since this is equally a historical myth, debunked in length by Amundson (2005). In addition, some even find population thinking in Aristotle: I take Lennox's (1987) claim that Aristotle was not after definitional essences of species as fixed entities as a claim that population thinking was nascent already in Aristotle. See also Balme (1980) and Pellegrin (1986). The philosophical tradition descends from Quine (1969). See, for instance, Matthews (1990) and Robertson (2013) on its history. See Charles (2000) for a book-length analysis of Aristotle's essentialism compared to contemporary points of views. Yet despite these contributions, it seems that the overall and myth-free history of essentialism has still to be written.

3. See R. A. Richards (2010), as a book length treatment on what species are.

4. Even the connection to essentialism might be historically contingent and is a matter of dispute: McOuat (2009) claims, for instance, that thinking in terms of natural kinds is not essentialist; rather, it replaced essentialist thinking. Naturally, if kind terms are at issue, linguistic issues arise too. Since this study ignores the linguistic debate, see instead Ben-Yami (2001) for a contemporary take on the contested relationship between linguistic aspects of kind terms and essentialism. The linguistic debate goes back to Putnam (1975) and Kripke (1972) and mainly concerns whether kind terms are descriptive (i.e., pointing at properties; denied by Ben-Yami, for

instance) and, if so, whether it is essential microstructural properties (in the sense of Putnam or Kripke) that fix membership in a kind or some other properties.

5. Similar pragmatic solutions can be found in Kitcher (1984), Dupré (1993), and Okasha (2002).

6. See Needham (2008) for the revisionary claim that (contrary to biological praxis) isotopes should be classified as separate substances (i.e., as separate kinds). I regard it as likely that ultimately more variation in chemical kinds will be discovered and that essentialism will break down even for chemical kinds, irrespective of whether Needham in particular is correct or not.

7. I owe this example to my students in my 2015 human nature class at Central European University. In discussing a first draft of this chapter, they came up with a first version of this thought example.

8. See Kronfeldner (2007, 497–498, 2011, 15) for more on Lamarckian evolution as typological in the just described sense and for distinguishing variational and transformational evolution.

9. Ferenc Huoranszki reminded me (personal communication) that in the past, as here, rationality was usually wedded to animality when humans were at issue. Like Aristotle, who specified humans as rational animals and not just as rational beings, the great "chain of being" of eighteenth-century philosophy had humans placed below higher spirits, such as angels or God, but higher than other animals and that meant they clearly were animals, made of flesh. "Man," Alexander Pope famously wrote in his poem *An Essay on Man*, "hangs between" in an "isthmus of a middle state"; a human has thus been perceived as a "strange hybrid monster," unique in being made of flesh *and* spirit, as Lovejoy (1936, 199) summarizes the eighteenth-century discussions about humans as part of the great chain of being. Given contemporary antinaturalism in philosophical anthropology and the philosophy of mind, and the current prevalence of discussions about artificial intelligence and enhancement, I take it that there are now philosophers (and maybe even plenty of them) who would not regard animality as necessary for humans to be humans. McMahan (2002), for instance, assumes that a human being is not (i.e., is not essentially) a biological organism but a "mind." Consequently, traits like rationality, intelligence, and the like are often introduced as characteristic of being human—without any reference to animality. Ultimately, though, I would argue, McMahan and all others who leave out animality have to admit that they talk about persons, not humans.

10. Imagine a field guide for twin world creatures visiting Earth, something similar to what people have when they go birdwatching.

11. See Kitcher (1999, 74), or Kittay (2008) for more depth in describing such examples.

12. For a defense of this claim, see Ereshefsky (2010b); for a critique, see Dumsday (2012). Buller (2005, 441–442), also stressed that for contemporary nonessentialist species concepts, two populations that are qualitatively exactly alike but without any genealogical connection would not count as one species.

13. The superchimp is a chimpanzee that has the cognitive capacities of a ten-year-old, at least in McMahan (2002).

14. Walsh as well as Lennox distinguish what they call "teleological essentialism," the kind of essentialism they think Aristotle defended, from "typological essentialism," treating the latter to refer to the first tenet of kind essentialism mentioned above. That means the latter refers to classification in general, or a fixed hierarchical taxonomy in particular. I am using the terms *taxonomic* or *classificatory essence* for what they call typological essentialism and contrast it with explanatory essence or essentialism, which can be teleological or not. The reason I made a different terminological choice is that both a classificatory and an explanatory essence can be typological in the sense discussed in this chapter.

15. A solution that Witt (2005) seems to favor and that I also take Lennox (1987) to adopt.

16. I would like to stay neutral with respect to whether capacities, potentialities, and dispositions are the same. I mainly use the term *capacity* since it is often used with respect to human beings and their status as humans (see, e.g., Jaworska and Tannenbaum 2013).

17. Thanks to Alexander Reutlinger for reminding me about this.

18. Humans are currently believed to share around 99 percent of their genes with their closest evolutionary relatives, despite the huge differences in life form. Pääbo (2001) is an entry point to the literature and the whereabouts and role of such percentages.

19. I take the distinction between ignoring and mistreating variation from a discussion with Christopher DiTeresi, who relates it to Love's (2009) suggestion to move the discussion about typological thinking away from metaphysics and toward an epistemology of representation that takes scientific practices more seriously, a move that I can only endorse.

20. It is a deep and, I think, contemporarily unsolved controversial issue whether and who has to adapt if there is a conflict. Must metaphysics correct its ontologies if there is a conflict with a scientific ontology, or must scientific ontology adapt to metaphysics, or can the two stay apart as incompatible? This question cannot be solved within the scope of this book, and I take the concrete claims developed here to not depend on such a solution.

Notes

4 The Developmentalist Challenge

1. A few parts of this chapter draw from Kronfeldner (2015). Permission for reuse is granted by Springer International Publishing Switzerland (© Springer International Publishing Switzerland 2015).

2. See also Zirkle (1946) on the long history of the belief in inheritance of acquired characteristics.

3. See Fancher (1983) on this episode of the history of the nature–nurture divide.

4. Note that when Galton introduced the divide between nature and nurture in his *English Men of Science* he did not use twin studies. He simply compared the rate of eminent men within families with the rate of those in the general population to find out what runs in families. Heritability obtained in what had a significantly higher frequency within families (higher similarity within family) compared to the general population.

5. I intentionally resist calling it a "multiple realizability relation" or the cultural level "emergent," which might involve again a hierarchy between the to-be-realized something (fundamental) and its realizers (emergent). I do not want to commit to such hierarchy between nature and culture, yet there are complex relations.

6. What I am offering here is a philosophical interpretation of Kroeber. For a more historical analysis, see Kronfeldner (2009) and Jackson (2010).

7. See Schaffner (1998) for a direct critique of developmental systems theory. Parity of causes, nonpreformationism, contextualism, indivisibility, and unpredictability are the assumptions he mentions and criticizes. See also Griesemer et al. (2005) and Merlin (2010).

8. See Jablonka and Raz (2009) or Bonduriansky (2012), focused on the biological debate; see Lock (2013), Meloni (2015, 2016), and Richardson and Stevens (2015) as an introduction to the literature discussing epigenetics with respect to the social sciences. The narrow sense used here goes back to Nanney (1958) and is also used, for instance, by Jablonka (2013, 76). Nicoglou and Merlin (2017) distinguish five meanings in five different epistemic contexts. Since the details do not matter here, I will only justify my use with respect to one alternative, a much broader sense of epigenetic inheritance that includes all kinds of effects of developmental resources as epigenetic, as in Griffiths and Stotz (2013, 112). Such a broad sense is, first, not big news (since cultural evolution, acknowledged since Weismann, would then simply be revamped as epigenetic), but also, second, a loss of differentiation that has no justification and, third, continues to boost the self-acclaimed enemy (any form of priority of genes) in order to fight it: it has genetic inheritance on the one hand and all the rest bundled together as "exogenetic" or "epigenetic" (broad sense). There is no need for such a gene-centered division of channels anymore, which is

why I opt for a more specific and balanced terminology of multiple inheritance channels.

9. Norms-of-reaction studies were introduced by Richard Woltereck. See Sarkar (1999) and Falk (2001) on the history of using and interpreting norm-of-reaction graphs.

10. Mackie (1974) called what I refer to as a "causal structure" a "causal field." The term *candidate gene* has become widespread in contemporary behavioral genetics literature.

11. See more on the history and philosophical intricacies of a quantitative apportioning of causes regarding ontogenetic development (how much is due to nature or nurture) in Sober ([1988] 1994). Sober gives a particularly lucid and seminal introduction on how the so-called analysis of variance (ANOVA) works, widespread in biological and social sciences and criticized famously by Lewontin ([1974] 2006). He also distinguishes the "how-much question" from claims about nature or nurture being more important than the other (i.e., calculating main effects and comparing them quantitatively), distinctions that would have been useful in Longino's (2013) treatment of incommensurability in nature-nurture debates. The literature on heritability is huge. Lewontin ([1974] 2006) gives a seminal critique of it. For an up-to-date overview, see Schaffner (2016), who outlines all the complexities involved in heritability measures and inferring causal importance from it, including the "missing heritability" problem: many of the behavioral geneticist's statistically established high heritabilities for traits (e.g., for body height), are missing in the molecular studies; they simply cannot find them.

12. For a contemporary exposition of how difference making connects to causation and explanation, see Woodward (2003). Only some accounts refer to counterfactuals.

13. This is the solution to the incommensurability problem that Longino (2013) unfortunately ignored.

14. See again Schaffner (2016) for a detailed exposition of such technical vocabulary.

15. See Paul and Brosco (2013) on the history of discovering the causal complexities involved in PKU.

16. Sometimes nonlinearity is also called "nonadditivity." Since the latter can also be understood as noncommensurability, I have chosen to use the term *nonlinearity*. As Sandra Mitchell (personal communication) stressed, nonlinearity does not mean that there cannot be a mathematical representation of the curve.

17. See, for instance, Lewontin ([1974] 2006) for further such patterns, or Kitcher (2001a, 398–399), for four different "deterministic schemes."

18. For a review and philosophical analysis of the debate, see Tabery (2014).

19. A point brought home famously by Lewontin ([1974] 2006), buffered by Sober ([1988] 1994), but clearly older, since Dobzhansky (1955) was already using it. Examination of the deeper history of that argument on nonexportability, however, has to wait for another occasion.

20. As a consequence, Sober ([1988] 1994) called the evaluation of difference making in matters of ontogenetic development *nonlocal*, that is, dependent on choice of a reference class, matters of fact extrinsic to the properties of the individuals entering the overall population-level considerations and in that sense nonlocal. His argument is basically the same as the one used by Lewontin ([1974] 2006, 2), who claimed that "the analysis [of variance] has a historical (i.e., spatiotemporal) limitation." This is why Lewontin calls it a "local analysis," local since it is contingent, depending on the sample of individuals and environments chosen.

21. The term *adaptation* is used here in the sense that behavioral ecology uses the term for an explanation of the origin of variation at the ontogenetic level, that is, as an ontogenetic process through which an organism changes in an environment and, in case of parental effects, even in response to the environment. Thanks to Mathieu Charbonneau for pushing me to clarify this, given that adaptation is often used in a different sense, that is, for an evolutionary process across individuals that is not necessarily providing an origin explanation.

22. See Robert (2008) and essays in Pigliucci and Müller (2010) for details of that debate.

23. For a review of the complexities in that debate with respect to conceptualizing inheritance, see Lamm (2012). See Kronfeldner (2011) on how Dawkins brings in culture via his concept of memes. Evolutionary psychologists such as Tooby and Cosmides (1992) are among those who tend to discount culture for evolutionary issues. For a direct critique of their approach, see Kronfeldner (2008).

24. As explained earlier in this chapter, a narrower and different, technical sense of heritability tries to find out how much of the similarity is due to biologically inherited resources alone. But this is not at issue here.

25. This is why I think Scott-Phillips, Dickins, and West's (2011) attack on the idea of cultural inheritance is misguided. They argue that culture is not of evolutionary importance since cultural inheritance is not an alternative to natural selection. I do not think that anyone defending the evolutionary importance of multiple inheritance systems would claim that cultural inheritance is an alternative to selection. All that is claimed is that if biological inheritance is evolutionarily important, then cultural inheritance is too, as part of the selection processes going on, not as alternative. Both can make a difference to selection and account for some stability and variation that occurs, without playing the same role as selection.

26. Cavalli-Sforza and Feldman (1981), Boyd and Richerson (1985), Richerson and Boyd (2005), and Durham (1991) developed such coevolutionary theories, or dual inheritance theories as they are also called, theories of cultural and biological change over long evolutionary times. The literature on this field has exploded in the last couple of years. It finally was widened toward multiple inheritance views in Jablonka and Lamb (2005), claiming that we have at least four different systems of heredity interacting in the evolution of organisms: genetic, epigenetic, behavioral, and symbolic heredity. A similar tradition stems from Sperber (1996), but with less reliance on transmission as important in explaining change in culture. Lactose tolerance is a standard example for coevolution. Aoki (1986) and Feldman and Cavalli-Sforza (1989) are seminal references to it. See Gerbault et al. (2011) for the complexities involved in modeling even this very plausible case of coevolution.

27. For a general analysis of their account and reply in line with what I develop in this book, see Griesemer et al. (2005) and Merlin (2010). I will not analyze their claims in general; I will critically answer only to the specific channelism-dependence point in chapter 5.

5 Genealogy, the Classificatory Nature, and Channels of Inheritance

1. This question could be further divided into questions for different kinds of explanations (e.g., so-called ultimate or proximate explanations), but to do so is not necessary for the aims at issue in this chapter. In addition, the distinction is controversial. See chapter 7 instead, where I will briefly discuss it.

2. Devitt (2008) is the most outspoken defender of species essentialism and is thus in focus in this chapter. But even those critical about essentialism leave the traditional phrase "what it is to be" rather unpacked (e.g., Haslanger 2014).

3. See Queiroz (2007) for a review of the discussion in taxonomy generally and Proctor 2003 for a review of the debates in paleoanthropology.

4. Thus, I agree with Dupré (1993, 51): "Nothing in evolutionary theory guarantees that genealogy will always provide the distinctions we need in order to understand the current *products* of evolution as opposed to the process by which they came to be." I disagree with him regarding genealogy deciding the issue for those evolutionary contexts where we would apply the concept of separate species (given that at the beginning of life it might simply not apply). Note that the claim that species are always evolutionary units (rather than a set of organisms without a lineage) does not mean that they are the only evolutionary units of selection (in contrast to genes or organisms).

5. Certainly, since the homeostatic mechanisms are not species specific, they cannot directly account for the species question in application (i.e., for how to delimit the boundaries of a particular species). Ereshefsky (2010b, 677) and Reydon

(2012, 235) claim that this is one of the reasons why Boyd's account of homeostatic mechanisms must fail.

6. The Preamble to the UN Declaration speaks of the "equal and inalienable rights of all members of the human family." Article 1 then says: "All human beings are born free and equal in dignity and rights. They are endowed with reason and conscience and should act towards one another in a spirit of brotherhood." "They" refers to "all human beings" which are identified only implicitly by "all members of the human family," which I interpret genealogically. Müller (2015) states that also other contemporary declarations of human rights include not much descriptive information about what it means to be human. I take this to confirm that genealogy has been key for contemporary declarations of human rights. Whether genealogy was also important in the deeper history of human rights is a much more complex issue that cannot be answered here.

7. This also answers Dupré (1993, 57), who writes: "Any sorting principle based on ancestry supposes that it is possible in some way to pick out the appropriate set of ancestors. In other words, being descended from one of the members of a particular set is no criterion at all unless there is some way of picking out the members of that set. ... Phylogeny, in short, cannot possibly create essences *ex nihilo*." They cannot, as I admitted, since a speciation event needs to be specified to decide the extension of the kind at the populational level (i.e., to delineate the chunks on the tree of life). If consequently the species question is solved, the partaking question becomes solvable too, and by genealogy alone. If the species question is not solved, you certainly cannot answer the partaking question since the question simply makes no sense.

8. See Okasha (2002, 203) for a similar claim. Since the answer to the partaking question rests on genealogy, this is compatible but in no way dependent on Kripke's idea that individuals have their origins as essence. Okasha (2002) gives some details on how Kripke fits the relationalist picture.

9. The qualifier *regularly* is important since the claim that biological reproduction involves material overlap is contested. Yet nothing in the literature discussing biological reproduction in detail (e.g., Griesemer 2000; Godfrey-Smith 2009; Merlin 2017) conflicts with the claim that biological reproduction and social learning are simply not the same kinds of causal interactions between people.

10. This is unlikely to change even if one were to distinguish different ways of how to count causal interactions. I thus assume near-decomposability to be a robust pattern across different ways of counting causal interactions. Thanks to Martin Kusch for pushing me on this point.

11. "Content" in the biological case is here interpreted functionally as "what the resource does for the organism."

12. Viruses are the exception for this claim about fixed modes of biological inheritance. As Isabella Sarto-Jackson (personal communication) remarked, with reference

to Pellett et al. (2012), genetic factors can sometimes switch modes of inheritance. For instance, one of the herpes viruses transmitted between humans can be transmitted vertically (by biological reproduction, from a parent to descendants) as well as horizontally to another adult, and even at the same time. But note that in order to really be analogous to the cultural case, the switch needs to happen in response to new evolutionary affordances, and it needs to be reversible to count as equally content dependent. It goes beyond the scope of this to determine whether that applies to the case of viruses.

13. The argument defended here is thus also consistent with Charbonneau's (draft, June 21, 2017) claim that there are no general ways to assess the degree of fidelity of a social transmission mechanism. Yet he goes beyond the issues discussed here and shows that assessing culture's fidelity (and therefore also the stability of culture, which is often taken to be in part at least explained by fidelity of specific transmission processes) is dependent on similarity judgments that are relative to the perspectives and epistemic interests of the scientists assessing the similarity. He thus offers a further kind of content dependence of cultural stability but one that is at the meta-level of an observer judging the stability by judging similarity. Claidière and André (2012) talked about the level of the individuals who learn from each other and thereby select a mode of inheritance as well as a direction of change, depending on the content and affordances.

14. The figure has one problem, though: it classifies parental effects as cases of inheritance, which they are not, as illustrated in section 4.3.

15. Although Griffiths and Gray (1994, 2001) can count as attacking the distinction of channels of inheritance, I take Griffiths and Gray (2004) as more channelism friendly since they stopped opposing it explicitly. They rather seem to use some concept of channels (and implicitly used it already while criticizing it in 2001) by referring to "extragenetic inheritance" in contrast to "genetic inheritance."

16. See Kelly and Oldring's (2015) CBC Radio feature "This Is That." Thanks to Gábor Betegh for drawing my attention to this example.

6   Toward a Descriptive Human Nature

1. See also Levins and Lewontin (1985), who show that standard assessments of typicality depend on social position and the like. Reydon (2014, 35) also claims that arguments from empirical studies to human nature (e.g., in the sense of typicality) are "hard to substantiate." Barrett (2017, 196) mentions that "we continually run the risk of misrepresenting how other people think and feel." She agrees, though, that anthropology can arrive at some knowledge about the historically changing life form of humans in general. Even one of the most outspoken defenders of "universals," Brown (2004), admits that it is hard to establish them. Brown is known for stressing that there are many generalizations about human life (e.g., mentioning

over three hundred in his 1991 book), many of which will be contested since, as he admits, there are "severe methodological limitations on what can be known about universals in general" (Brown 2004, 49). He mentions, for instance, sampling of people or that descriptions stem from different studies, with different definitions of the phenomena or different methodological assumptions used. See also Gough (2014) or Antweiler (2016) for a detailed account of the complexity that goes into establishing universal claims about the human life form.

2. I take Machery (2008), Stotz (2010), Griffiths (2011), Stotz and Griffiths (forthcoming), and Ramsey (2013, forthcoming) as defending such a descriptivist concept of human nature as the best replacement candidate for the outdated essentialist human nature. Hull, Dupré (2011), Lewens (2012), and Downes (2016) would also agree that some kind of descriptive nature exists. Still, they disagree with the other authors on details, and at least Downes (2016) opts for the irrelevance of generalizations about humans and thus for the irrelevance of the concept of a descriptive "human nature."

3. Each property of the cluster can be but is not necessarily a property of individuals. They can be properties of specific time slices of individuals or of groups of individuals: there are typical properties that apply only to a specific time slice of human beings (e.g., properties that only children have) and there are typical properties that are strictly seen not intrinsic properties of individuals but relational properties (e.g., certain kinds of family bonds). See Machery (forthcoming) on this important clarification.

4. As an exception, see Antony (2000, 34).

5. The distinction between numerical (being one, who one is) and qualitative identity (how one is) is widespread in discussions about personal identity. I am using it here to indicate that there are some interesting parallels between discussing personal identity and discussing human nature. A direct comparison of these parallels, however, has to wait for another occasion.

6. Only if a species-specific trait is treated as a mere indicator of membership in the kind, as in a field guide, it is not assumed to be simultaneously typical.

7. Ekman (1993) is the twentieth-century authority for confirming Darwin's claims on the expression of emotions.

8. Downes (2016) seems to side with Machery with respect to the disjunctive method and is critical of the usefulness of generalizations about humans in general.

9. Ramsey (forthcoming) stresses that his account is about robust relationships between traits rather than a "trait bin account" of human nature (i.e., an account about traits that are put into an epistemic bin that is then called "human nature"). Yet his account is still putting things in a bin, namely, a bin of what he calls "trait clusters" or trait-trait "associations." What he means by that is simply what others

(e.g., Kitcher and me; see chapter 4 above) have called developmental dependencies between certain developmental factors and the resulting phenotypic traits at issue. The causal relation between the disease PKU (as introduced in chapter 4), the gene for PAH, and a certain diet becomes part of human nature in Ramsey's all-inclusive account since that causal relation is a robust relation between the "consequent" trait standardly called PKU and the luckily not very prevalent "antecedent" trait "lacking PAH + consuming phenylalanine" (a trait, in Ramsey's account, not a combination of internal and external causal factors, as standard). This might be a new way of talking, but otherwise it is still simply an account with a greatly enlarged bin of trait-trait-associations resulting from developmental resources, a very inclusive bin. Finally, there is already a way of including Ramsey's trait-trait associations: the way traditionally done, as "conditional universals" of the human life form, as in Brown (2004).

10. During a discussion in December 2015 at Cambridge University, one line of argument regarding establishing typicality for specific traits was that it must be empirically determined when the abstraction is adequate and when not. Yet nothing at all was said, by the philosophers or the scientists present, how adequacy of abstraction should be determined.

11. One can even show that those opposing the nature-nurture divide sometimes simply hide it behind a holistic veneer. Ramsey (2013, forthcoming) is such a case. It is hidden behind the matrix of the life history trait cluster because intervening on genetic factors is treated as changing the nature of an individual, whereas "varying the way that an individual encounters its environmental heterogeneity" only "reveals something about its nature" rather than changing it (forthcoming, 49).

12. This paragraph has been taken more or less verbatim from Kronfeldner et al. (2014), with permission from my coauthors, Neil Roughley and Georg Toepfer.

7   The Stability of Human Nature

1. The literature on the topic is vast. For a recent overview, see Khalidi (2016).

2. The simplest specification is undoubtedly to regard "innate" as denoting solely what the word indicates: "in-nate," referring to a temporal distinction only—prenatal versus postnatal, as it did according to Keller (2010) in the nineteenth century.

3. Ignoring this evolutionary dimension is one of the reasons Ramsey's (2013, forthcoming) account and Stotz and Griffiths (forthcoming) are too inclusive.

4. See his replies on the connection between innateness and evolvedness in a recent interview (Machery 2016b, 60–62).

5. See Oyama (1985) or Griffiths and Gray (1994). Laimann (2014) has also criticized Machery along these lines since ultimate (why X got selected) and proximate explanations (how X works and develops) are complementary, not alternative.

6. I take the concept of scaffolding from Wimsatt and Griesemer (2007).

7. Smith (2012), building on Machery (2008) in using *evolved* as a qualifier for what is part of human nature, explicitly assumes that evolution is possible only if based on genetic inheritance. This is clearly wrong.

8. Griffiths and Gray (2004, 412) also mention a case from ant species, where "a 'mutation' in a nongenetic element of the developmental matrix can induce a new self-replicating variant of the system that may differ in fitness from the original."

9. See, as mentioned earlier, Enard et al. (2002) as an entry to the literature.

10. See Kronfeldner (2015) for more details on reconstituting phenomena via abstraction from details.

11. Thanks to Jessica Laimann for pushing me to clarify this point.

8 An Explanatory Nature

1. See, for instance, Reydon (2009, 726) on essences.

2. Boulter (2012) has a similar approach, based on intrinsic to the organism "developmental programmes" (more than genes, but still intrinsic). Here, I mainly reply to Devitt's account and reply to one specific point from Boulter later in the chapter.

3. Personal communication.

4. See Reutlinger, Schurz, and Hüttemann (2011) on lazy ceteris paribus clauses.

9 Causal Selection and How Human Nature Is Thereby Made

1. Parts of this chapter relate to major sections of Kronfeldner (2014), albeit in a different and more detailed form. Reuse acknowledged by Oxford University Press. (© Maria Kronfeldner, 2014.)

2. There are a couple of terms for this in the literature: *Multifactoriality* is a term mainly used by scientists themselves; there is also the term *multicausality*. Menzies (2004) uses the term *problem of profligate causes*. *Causal complexity* is another term in use, although it has a broader meaning: it can also refer to, for example, one cause having many effects (in addition to one effect having many causes), nonlinear dynamics, or emergent phenomena.

3. For arguments toward the necessary partiality of our explanations independent of causal selection, see, for instance, van Fraassen (1980) or Mitchell (2003, 2009).

4. Certainly, at a certain point and if possible, scientists integrate their partial explanations with other partial explanations from other scientists, as discussed in Kronfeldner (2015).

5. Schaffner (2006) builds on this as well when he talks about field elements and preferred causal model systems.

6. A term I take from Dietl (1970). Less well known is that even Mill (1858, Book 3, ch. 5, §3) can be interpreted as having pointed to the normality criterion when he distinguishes causes that are events (extraordinary, coming last into existence) from causes that are states of affairs (ordinary, permanent). Hitchcock and Knobe (2009) backed up Hart and Honoré's ([1959] 1985) seminal normality argument (which relied on ordinary language philosophy) with evidence from so-called experimental philosophy: they experimentally tested how subjects judge, if asked "What is the cause?" (or a similarly suitable question about "What if things had been different?"). The results are in favor of the normality principle: people generally use a/normality as guide in their causal judgments, be it in social or biological contexts. Hitchcock and Knobe also added an explication and defense of that tendency in terms of relevant versus irrelevant counterfactuals entering causal judgments. Their paper is a superb amalgamation of empirical data and philosophical reasoning, but the additions they provide are not important in this context, since the basic point, as they admit (601), is still the one from Hart and Honoré. We select the abnormal causal factor as the one we use in our causal judgments and regard the others as mere background factors.

7. Hart and Honoré ([1959] 1985) were explicit about that requirement. Whether it holds is important. Weber (forthcoming), for instance, also argues that normality guides us in our causal selection, but he has it upside down: he assumes that scientists want to learn about the mechanistic details of how things normally work. They thus foreground what is normal and background what is anormal. This is indeed the case, but it holds only for contexts where explanations are mechanistic explanations of how things run normal. It will not hold for what Collingwood called practical sciences. Weber (personal communication) agrees that how normality influences causal selection depends on the context of inquiry, that is, on what is typically at issue in a field.

8. I take the term *context of occurrence* from Menzies (2004).

9. This argument holds even if somewhere down the road, the two causal routes converge at a more direct cause of the effect (e.g., in the brain, where certain hormones are produced). As long as there are two different roads to change that more direct cause, there are two different distal causes. Thanks to Dennis Lehmkuhl and Eleanor Knox for pushing me on that point.

10. Hart and Honoré ([1959] 1985) brought the point home with respect to cancer, an example Collingwood did not treat with enough care. For a more general critique of actual controllability see, e.g., Hausman (1986, 1997) and Menzies and Price (1993).

11. Inability to control can, however, psychologically constrain willingness to control. If there is no hope whatsoever that we can ever gain positive control over the

causal factor, it is rather unreasonable to be willing to control it, as Alexander Reutlinger has stressed in response to my approach (personal communication). His example was that it would be unreasonable to be willing to control the half-live of uranium since we cannot physically intervene on it. See Reutlinger (2012) for details on this example and the role of interventions in causal explanations more generally.

12. So far, the molecularization of life has been explained in a slightly different instrumentalist manner. The instrumentalist claim is that genes have been so often prioritized in explanations of human traits because in experimental contexts, they are better instrumental handles, that is, technologically more tractable than environmental factors (see, e.g., Kitcher 1996, 2001a; Schaffner 1998; Gannett 1999; Keller 2000). While we can screen genes and experimentally intervene in precise ways on genes, we cannot screen or intervene on the environment in a similar technologically easy way. Genes are tractable, and the environment is not. In the analytic terms used in this study, instrumentalists assume that geneticists have better control (forward and backward) over genes than over the environment and therefore focus on genes as explainers. Although this instrumentalist approach to prioritizing genes can be directly derived from Collingwood's control principle (and Collingwood has been mentioned with respect to it), it ignores the issues that point to willingness to control as a revised version of Collingwood. The ability to manipulate cannot explain the prioritizing that we saw for roughly the first three-quarters of the twentieth century and it cannot explain the prioritizing between equally controllable factors, that is, in cases where having the means to control something still leaves room for a choice.

13. For a concept of classificatory looping effects that is broader, not requiring reflexivity, see Khalidi (2010). Or imagine a case of reciprocal or feedback causation, for example, in gene regulation, where genes are causally involved in the production of proteins that are causally involved in the production of genes. Carrier and Finzer (2006) call such a reciprocal causal feedback an "explanatory loop." In my use of the term, this would not be a looping effect in the narrow sense, since there is no reflexivity of reacting to the explanation involved, even though any feedback process is certainly a loop. Yet despite reflexivity being crucial, a looping effect does not have to involve a conscious reaction to the explanation. It is sufficient for reflexivity if the population reacts to the explanation and changes itself and its explanations thereby. Thanks to Alexander Reutlinger for pushing me on this point.

14. The matching between explanatory and descriptive level will certainly not be simple and clear-cut, as can be derived from the interactionist consensus.

10 Humanism and Normativity

1. Ryder introduced the term *speciesism* in 1970 in a leaflet. See Ryder (2010). For a defense of the assumption that genealogy as such is morally irrelevant, see McMahan (2002, 209–228).

2. For review of the debates, see Wasserman et al. (2013).

3. A terrorist has proven to not interact with other humans in morally adequate ways. Yet according to the double-entry solution, the terrorist would still have to be granted human rights. Thanks to Mara-Daria Cojocaru for pushing me to make this case clear. The eremite, a case I owe to Jessica Laimann in response to the double-entry solution, is the exact opposite: it had no chance of showing whether she or he would interact in morally adequate ways, but it has moral standing as long as we can reasonably expect that individual to stand in either genealogical or social relations with other humans. So neither case poses a problem for the double-entry solution.

4. See McMahan (2002), Kittay (2008), Kittay and Carlson (2010), and Wasserman et al. (2013).

5. This is also why Nussbaum's account has significant similarities to the theory of universal human needs developed by Doyal and Gough (1991), who base their approach on a neo-Kantian argument to establish two basic needs, health and autonomy, that are taken to be constitutive of all other universal needs. For the latter, they use "codified knowledge of the natural and social sciences" for a list of universal needs and eleven universal satisfiers of these needs (e.g., nutritional food and clean water, a secure childhood). The respective knowledge about these features of the descriptive human nature is based on cross-cultural data and taken to develop in interaction with the people concerned (Gough 2014, 368). Thanks to Michael Dover for recommending Doyal and Gough's theory for comparison.

6. In addition, it is also not specific to humans: consequently, one would have to grant any moral obligation or right based on that basic need to lots of other creatures too (and not just animals) if one does not want to fall back to speciesism.

11 Should We Eliminate the Language of Human Nature?

1. See Gayon (2003, 2008), Brattain (2007), Gissis (2008), Ludwig (2014), or Berg, Schor, and Soto (2014).

2. See Wilson (2018) or Wilson and St. Pierre (2016) on eugenics and Wolbring (2008) and Kittay and Carlson (2010) on disability.

3. I take the term *human prejudice* from B. Williams (2006).

4. This does not mean that it is ethically irrelevant to think about the rights of other animals (e.g., in terms of protection from suffering). But this book is not an attempt to deal with the issue of animal rights (i.e., whether only humans have certain rights). It rather deals with whether all humans have the same rights (the dehumanization problem). The issues are certainly connected but distinct enough to justify treatment of the one without the other to answer the questions this book asks.

Furthermore, I do not think that discussions about animal rights should be understood to literally negotiate the human-animal boundary in the sense of negotiating whether humans are genealogically (i.e., reproductively) separate from other animals. Even Haraway (2004, 2) agrees that humans and animals are separate species.

5. I take Smith (2011, 2014) to claim that dehumanization requires observing an individual as equally human and denying a hidden essence. This assumes psychological essentialism (not directly at issue in this study) rather than scientific essentialism. Yet even psychological essentialism is not necessary for dehumanization to occur. Although there are such cases where psychological essentialism explains the dehumanization happening, I would argue for two things. Even without systematic empirical studies, it is clear that not all cases of dehumanization involve an observable–nonobservable divide. If women are dehumanized as less rational (one of the properties that has historically been used for their dehumanization), they are neither perceived to be the same but assumed to be so (i.e., of the same genealogy, descending as men from human parents), nor was rationality standardly conceived as an unobservable hidden essence. In addition, if dehumanization does involve an observably-equal-but-not-in-essence dehumanization, that essence that is denied might simply be the genealogy itself. Psychological essentialism is not necessarily based on belief in an intrinsic essence (such as "having a soul," an example Smith uses). Genealogy can serve the purpose of an unobservable (albeit classificatory only) essence equally well, even today. But I admit that the connection between psychological essentialism and dehumanization needs further systematic study that has to wait for another occasion.

6. I do not think that the latter, claiming that the statistical generalization holds, conflicts with claims about the historical constructedness of homosexuality and heterosexuality as identities. See Bondarenko (2017) for details on this topic.

References

Abulafia, David. 2008. *The Discovery of Mankind: Atlantic Encounters in the Age of Columbus.* New Haven, CT: Yale University Press.

Amundson, Ron. 2000. Against normal function. *Studies in History and Philosophy of Science Part C: Studies in History and Philosophy of Biological and Biomedical Sciences* 31 (1): 33–53.

Amundson, Ron. 2005. *The Changing Role of the Embryo in Evolutionary Thought: Roots of Evo-Devo.* Cambridge: Cambridge University Press.

Antony, Louise M. 1998. "Human nature" and its role in feminist theory. In *Philosophy in a Feminist Voice: Critiques and Reconstructions*, edited by Janet A. Kourany, 63–91. Princeton, NJ: Princeton University Press.

Antony, Louise M. 2000. Natures and norms. *Ethics* 111 (1): 8–36.

Antweiler, Christoph. 2016. *Our Common Denominator: Human Universals Revisited.* New York: Berghahn.

Aoki, Kenichi. 1986. A stochastic model of gene-culture coevolution suggested by the "culture historical hypothesis" for the evolution of adult lactose absorption in humans. *Proceedings of the National Academy of Sciences of the United States of America* 83 (9): 2929–2933.

Aristotle. 1952. *The Works of Aristotle.* Edited by William D. Ross and John A. Smith. Oxford: Oxford University Press.

Ayala, Francisco J. 2016. Interview on human nature. In *Conversations on Human Nature*, edited by Agustín Fuentes and Aku Visala, 43–54. London: Routledge, Taylor and Francis.

Balme, David M. 1980. Aristotle's biology was not essentialist. *Archiv für Geschichte der Philosophie* 62 (1): 1–12.

Barrett, Louise. 2017. What is human nature (if it is anything at all?). In *Routledge Handbook of Evolution and Philosophy*, edited by Richard Joyce, 194–209. New York: Routledge.

Bayertz, Kurt. 2009. Hat der Mensch eine "Natur"? Und ist sie wertvoll? In *Bios und Zoë: Die menschliche Natur im Zeitalter ihrer technischen Reproduzierbarkeit*, edited by Martin G. Weiss, 191–218. Frankfurt am Main: Suhrkamp.

Bayertz, Kurt. 2013. *Der aufrechte Gang: Eine Geschichte des anthropologischen Denkens*. Munich: Beck.

Ben-Yami, Hanoch. 2001. The semantics of kind terms. *Philosophical Studies* 102 (2): 155–184.

Berg, Manfred, Paul Schor, and Isabel Soto. 2014. The weight of words: Writing about race in the United States and Europe. *American Historical Review* 119 (3): 800–808.

Bird, Alexander. 2007. *Nature's Metaphysics: Laws and Properties*. New York: Oxford University Press.

Birnbacher, Dieter. 2009. Wieweit lassen sich moralische Normen mit der "Natur des Menschen" begründen? In *Bios und Zoë: Die menschliche Natur im Zeitalter ihrer technischen Reproduzierbarkeit*, edited by Martin G. Weiss, 219–239. Frankfurt am Main: Suhrkamp.

Böhme, Gernot. 2002. On human nature. In *On Human Nature: Anthropological, Biological, and Philosophical Foundations*, edited by Armin Grunwald, Mathias Gutmann, and Eva M. Neumann-Held, 3–14. Berlin: Springer.

Bondarenko, Olesya. 2017. Two theories better than one? Integration and epistemic values in research on sexual orientation. MA thesis, Central European University, Budapest.

Bonduriansky, Russell. 2012. Rethinking heredity, again. *Trends in Ecology and Evolution* 27 (6): 330–336.

Boulter, Stephen J. 2012. Can evolutionary biology do without Aristotelian essentialism? *Royal Institute of Philosophy Supplements* 70:83–103.

Boyd, Richard. 1991. Realism, anti-foundationalism and the enthusiasm for natural kinds. *Philosophical Studies* 61:127–148.

Boyd, Richard. 1999. Homeostasis, species, and higher taxa. In *Species: New Interdisciplinary Essays*, edited by Robert A. Wilson, 141–185. Cambridge, MA: MIT Press.

Boyd, Robert, and Peter J. Richerson. 1985. *Culture and the Evolutionary Process*. Chicago: University of Chicago Press.

Brattain, Michelle. 2007. Race, racism, and antiracism: UNESCO and the politics of presenting science to the postwar public. *American Historical Review* 112 (5): 1386–1413.

Brown, Donald E. 1991. *Human Universals*. New York: McGraw-Hill.

Brown, Donald E. 2004. Human universals, human nature and human culture. *Daedalus* 133 (4): 47–54.

Buchanan, Allen E. 2011. *Beyond Humanity? The Ethics of Biomedical Enhancement.* Oxford: Oxford University Press.

Buller, David J. 2005. *Adapting Minds: Evolutionary Psychology and the Persistent Quest for Human Nature.* Cambridge, MA: MIT Press.

Carrier, Martin, and Patrick Finzer. 2006. Explanatory loops and the limits of genetic reductionism. *International Studies in the Philosophy of Science* 20 (3): 267–283.

Caspi, Avshalom, Joseph McClay, Terrie E. Moffitt, Jonathan Mill, Judy Martin, Ian W. Craig, Alan Taylor, and Richie Poulton. 2002. Role of genotype in the cycle of violence in maltreated children. *Science* 297 (5582): 851–854.

Cassirer, Ernst. 1944. *An Essay on Man: An Introduction to a Philosophy of Human Culture.* New Haven, CT: Yale University Press.

Cavalli-Sforza, Luigi L. 2000. *Genes, Peoples and Languages.* New York: North Point Press.

Cavalli-Sforza, Luigi L., and Marc Feldman. 1981. *Cultural Transmission and Evolution: A Quantitative Approach.* Princeton, NJ: Princeton University Press.

Charbonneau, Mathieu. 2017. Cultural fidelity. Unpublished manuscript.

Charles, David. 2000. *Aristotle on Meaning and Essence.* Oxford: Clarendon Press.

Claidière, Nicolas, and Jean-Baptiste André. 2012. The transmission of genes and culture: A questionable analogy. *Evolutionary Biology* 39 (1): 12–24.

Clark, William, Jan Golinski, and Simon Schaffer, eds. 1999. *The Sciences in Enlightened Europe.* Chicago: University of Chicago Press.

Collingwood, Robert G. 1940. Three senses of the word "cause." In *An Essay on Metaphysics*, 285–327. Oxford: Clarendon Press.

Collingwood, Robert G. 1945. *The Idea of Nature.* Oxford: Clarendon Press.

Corbey, Raymond. 2005. *The Metaphysics of Apes: Negotiating the Animal-Human Boundary.* Cambridge: Cambridge University Press.

Crane, Tim. 2013. Human uniqueness and the pursuit of knowledge. In *Contemporary Philosophical Naturalism and Its Implications*, edited by Bana Bashour and Hans Muller, 139–145. London: Routledge.

Danchin, Étienne, Anne Charmantier, Frances A. Champagne, Alex Mesoudi, Benoit Pujol, and Simon Blanchet. 2011. Beyond DNA: Integrating inclusive inheritance into an extended theory of evolution. *Nature Reviews. Genetics* 12 (7): 475–486.

Darwin, Charles. 1859. *On the Origin of Species by Means of Natural Selection, or the Preservation of Favoured Races in the Struggle for Life.* London: Murray.

Daston, Lorraine. Forthcoming. *Die Leidenschaften des Unnatürlichen*. Berlin: Matthes & Seitz.

Dawkins, Richard. 1976. *The Selfish Gene*. Oxford: Oxford University Press.

Dawkins, Richard. 1982. *The Extended Phenotype*. Oxford: Freeman.

Dennett, Daniel C. 1991. Real patterns. *Journal of Philosophy* 88 (1): 27–51.

Descola, Philippe. 2005. *Par-delà nature et culture*. Paris: Gallimard.

Deslauriers, Marguerite. 1998. Sex and essence in Aristotle's metaphysics and biology. In *Feminist Interpretations of Aristotle*, edited by Cynthia A. Freeland, 138–167. University Park: Pennsylvania State University Press.

Desmond, Adrian J., and James R. Moore. 2009. *Darwin's Sacred Cause: Race, Slavery and the Quest for Human Origins*. London: Penguin.

Devitt, Michael. 2008. Resurrecting biological essentialism. *Philosophy of Science* 75 (3): 344–382.

Dietl, Paul J. 1970. Abnormalism. *Theoria* 36 (2): 93–99.

Dobzhansky, Theodosius G. 1955. *Evolution, Genetics, and Man*. New York: Wiley.

Dobzhansky, Theodosius G. 1956. *The Biological Basis of Human Freedom*. New York: Columbia University.

Downes, Stephen M. 2016. Confronting variation in the social and behavioral sciences. *Philosophy of Science* 83 (5): 909–920.

Downes, Stephen, and Edouard Machery, eds. 2013. *Arguing about Human Nature: Contemporary Debates*. New York: Routledge.

Doyal, Len, and Ian Gough. 1991. *A Theory of Human Need*. New York: Guilford Press.

Dumsday, Travis. 2012. A new argument for intrinsic biological essentialism. *Philosophical Quarterly* 62 (248): 486–504.

Dupré, John. 1993. *The Disorder of Things: Metaphysical Foundations of the Disunity of Science*. Cambridge, MA: Harvard University Press.

Dupré, John. 2001. *Human Nature and the Limits of Science*. Oxford: Clarendon.

Dupré, John. 2011. What is natural about human nature? In *Lebenswelt Und Wissenschaft. I. Deutscher Kongreß Für Philosophie. Kolloquiumsbeiträge*, edited by C. F. Gethmann, 160–171. Hamburg: Meiner.

Dupré, John. Forthcoming. Human nature: A process perspective. In *Why We Disagree about Human Nature*, edited by Elisabeth Hannon and Tim Lewens, 92–107. Oxford: Oxford University Press.

Durham, William H. 1991. *Coevolution: Genes, Culture and Human Diversity*. Stanford, CA: Stanford University Press.

Ekman, Paul. 1993. Facial expression and emotion. *American Psychologist* 48 (4): 384–392.

Ellis, Brian. 2001. *Scientific Essentialism*. Cambridge: Cambridge University Press.

Enard, Wolfgang, Molly Przeworski, Simon E. Fisher, Cecilia S. L. Lai, Victor Wiebe, Takashi Kitano, Anthony P. Monaco, and Svante Pääbo. 2002. Molecular evolution of FOXP2, a gene involved in speech and language. *Nature* 418 (6900): 869–872.

Ereshefsky, Marc. 2010a. Species. In *The Stanford Encyclopedia of Philosophy*, edited by Edward N. Zalta. Metaphysics Research Lab, Stanford University. http://plato.stanford.edu/archives/spr2010/entries/species/.

Ereshefsky, Marc. 2010b. What's wrong with the new biological essentialism. *Philosophy of Science* 77 (5): 674–685.

Eyssel, Friederike, and Frank Hegel. 2012. (S)he's got the look: Gender stereotyping of robots. *Journal of Applied Social Psychology* 42 (9): 2213–2230.

Falk, Raphael. 2001. Can the norm of reaction save the gene concept? In *Thinking about Evolution: Historical, Philosophical, and Political Perspectives*, edited by Rama S. Singh, Costas B. Krimbas, Diane B. Paul, and John Beatty, 119–140. Cambridge: Cambridge University Press.

Fancher, Raymond E. 1983. Alphonse de Candolle, Francis Galton, and the early history of the nature–nurture controversy. *Journal of the History of the Behavioral Sciences* 19 (4): 341–352.

Feldman, Marcus W., and Luigi L. Cavalli-Sforza. 1989. On the theory of evolution under genetic and cultural transmission with application to the lactose absorption problem. In *Mathematical Evolutionary Theory*, edited by Marcus W. Feldman, 145–173. Princeton, NJ: Princeton University Press.

Finnis, John. 2011. *Natural Law and Natural Rights*. 2nd ed. Oxford: Oxford University Press.

Fish, Eric W., Dara Shahrokh, Rose Bagot, Christian Caldji, Timothy Bredy, Moshe Szyf, and Michael J. Meaney. 2004. Epigenetic programming of stress responses through variations in maternal care. *Annals of the New York Academy of Sciences* 1036 (1): 167–180.

Foot, Philippa. 2001. *Natural Goodness*. Oxford: Oxford University Press.

Francis, Gregory. 2013. Replication, statistical consistency, and publication bias. *Journal of Mathematical Psychology* 57 (5): 153–169.

Francis, Gregory. 2015. Excess success for three related papers on racial bias. *Frontiers in Psychology* 6 (512). doi:10.3389/fpsyg.2015.00512.

Fricker, Miranda. 2016. Epistemic injustice and the preservation of ignorance. In *The Epistemic Dimensions of Ignorance*, edited by Rik Peels and Martijn Blaauw, 160–177. Cambridge: Cambridge University Press.

Fukuyama, Francis. 2002. *Our Posthuman Future*. New York: Farrar, Straus and Giroux.

Gallie, Walter B. 1956. Essentially contested concepts. *Proceedings of the Aristotelian Society* 56 (March 12): 167–198.

Galton, Francis. 1874. *English Men of Science: Their Nature and Nurture*. London: Macmillan.

Galton, Francis. 1892. *Hereditary Genius: An Inquiry into Its Laws and Consequences*. 2nd ed. London: Macmillan.

Gannett, Lisa. 1999. What's in a cause? The pragmatic dimensions of genetic explanations. *Biology and Philosophy* 14 (3): 349–373.

Gannett, Lisa. 2008. Genes and society. In *The Oxford Handbook of Philosophy of Biology*, edited by Michael Ruse, 451–477. Oxford: Oxford University Press.

Gayon, Jean. 2003. Do the biologists need the expression "human race"? UNESCO 1950–1951. In *Bioethical and Ethical Issues Surrounding the Trials and Code of Nuremburg*, edited by Jacques Rozenberg, 23–48. New York: Edwin Melon Press.

Gayon, Jean. 2008. Is there a biological concept of race? *NTM—Zeitschrift für Geschichte der Wissenschaften, Technik und Medizin* 16 (3): 365–370.

Gebhardt, Wolfgang. 2008. Der Urknall und die chemischen Elemente: Auf den Spuren der chemischen Evolution. In *Evolution: 150 Jahre nach Darwin*, 17–46. Regensburg: Universitätsverlag Regensburg.

Gelman, Susan A. 2003. *The Essential Child: Origins of Essentialism in Everyday Thought*. Oxford: Oxford University Press.

Gerbault, Pascale, Anke Liebert, Yuval Itan, Adam Powell, Mathias Currat, Joachim Burger, Dallas M. Swallow, and Mark G. Thomas. 2011. Evolution of lactase persistence: An example of human niche construction. *Philosophical Transactions of the Royal Society of London B: Biological Sciences* 366 (1566): 863–877.

Gissis, Snait B. 2008. When is "race" a race? 1946–2003. *Studies in History and Philosophy of Science Part C: Studies in History and Philosophy of Biological and Biomedical Sciences* 39 (4): 437–450.

Godfrey-Smith, Peter. 2009. *Darwinian Populations and Natural Selection*. Oxford: Oxford University Press.

Goff, Phillip Atiba, Jennifer L. Eberhardt, Melissa J. Williams, and Matthew Christian Jackson. 2008. Not yet human: Implicit knowledge, historical dehumanization, and contemporary consequences. *Journal of Personality and Social Psychology* 94 (2): 292–306.

Goriely, Stephane, Andreas Bauswein, and Hans-Thomas Janka. 2011. R-process nucleosynthesis in dynamically ejected matter of neutron star mergers. *Astrophysical Journal: Letters* 738 (2): L32.

Gough, Ian. 2014. Lists and thresholds: Comparing the Doyal-Gough theory of human need with Nussbaum's capabilities approach. In *Capabilities, Gender, Equality towards Fundamental Entitlements*, edited by Flavio Comim and Martha Craven Nussbaum, 357–382. Cambridge: Cambridge University Press.

Gould, Stephen J. 1978. Women's brains. *New Scientist*, November 2, 364–366.

Griesemer, James R. 2000. Development, culture and the units of inheritance. *Philosophy of Science: Proceedings of the Biennial Meetings of the Philosophy of Science Association* 67 (S): S348–368.

Griesemer, James R., Matthew H. Haber, Grant Yamashita, and Lisa Gannett. 2005. Critical notice: Cycles of contingency—developmental systems and evolution. *Biology and Philosophy* 20 (2–3): 517–544.

Griffiths, Paul E. 1999. Squaring the circle: Natural kinds with historical essences. In *Species: New Interdisciplinary Essays*, edited by Robert A. Wilson, 209–228. Cambridge, MA: MIT Press.

Griffiths, Paul E. 2002. What is innateness? *Monist* 85 (1): 70–85.

Griffiths, Paul E. 2011. Our plastic nature. In *Transformations of Lamarckism: From Subtle Fluids to Molecular Biology*, edited by Snait B. Gissis and Eva Jablonka, 319–330. Cambridge, MA: MIT Press.

Griffiths, Paul E., and Russell D. Gray. 1994. Developmental systems and evolutionary explanation. *Journal of Philosophy* 91 (6): 277–304.

Griffiths, Paul E., and Russell D. Gray. 2001. Darwinism and developmental systems. In *Cycles of Contingency: Developmental Systems and Evolution*, edited by Susan Oyama, Paul Griffiths, and Russell D. Gray, 195–218. Cambridge, MA: MIT Press.

Griffiths, Paul E., and Russell D. Gray. 2004. The developmental systems perspective: Organism-environment systems as units of evolution. In *Phenotypic Integration: Studying the Ecology and Evolution of Complex Phenotypes*, edited by Massimo Pigliucci and Katherine Preston, 409–431. Oxford: Oxford University Press.

Griffiths, Paul E., and Karola Stotz. 2013. *Genetics and Philosophy: An Introduction*. Cambridge Introductions to Philosophy and Biology. Cambridge: Cambridge University Press.

Hacker, Peter. 2010. *Human Nature: The Categorical Framework*. Malden, MA: Wiley Blackwell.

Hacking, Ian. 1986. Making up people. In *Reconstructing Individualism: Autonomy, Individuality and the Self in Western Thought*, edited by Thomas Heller, Morton Sosna, and David Wellbery, 222–236. Stanford, CA: Stanford University Press.

Hacking, Ian. 1991. A tradition of natural kinds. *Philosophical Studies* 61 (1–2): 109–126.

Hacking, Ian. 1995. The looping effects of human kinds. In *Causal Cognition*, edited by Dan Sperber, David Premack, and Ann James Premack, 351–383. Oxford: Oxford University Press.

Hacking, Ian. 2007a. Kinds of people: Moving targets. *Proceedings of the British Academy* 151:285–318.

Hacking, Ian. 2007b. Natural kinds: Rosy dawn, scholastic twilight. *Royal Institute of Philosophy* 61 (Suppl.): 203–239.

Hannon, Elisabeth, and Tim Lewens. Forthcoming. *Why We Disagree about Human Nature*. Oxford: Oxford University Press.

Haraway, Donna J. 2004. *The Haraway Reader*. New York: Routledge.

Hart, Herbert L. A., and Tony Honoré. (1959) 1985. *Causation in the Law*. 2nd ed. Oxford: Clarendon Press.

Haslam, Nick. 2006. Dehumanization: An integrative review. *Personality and Social Psychology Review* 10 (3): 252–264.

Haslam, Nick, and Steve Loughnan. 2014. Dehumanization and infrahumanization. *Annual Review of Psychology* 65 (1): 399–423.

Haslanger, Sally. 2014. The normal, the natural and the good: Generics and ideology. *Politica & Società: Periodico di Filosofia Politica e Studi Sociali* 3 (3): 365–392.

Haste, Helen. 2000. Are women human? In *Being Human*, edited by N. Roughley, 175–196. Berlin: de Gruyter.

Hausman, Daniel M. 1986. Causation and experimentation. *American Philosophical Quarterly* 23 (2): 143–154.

Hausman, Daniel M. 1997. Causation, agency, and independence. *Philosophy of Science, Proceedings of the Biennial Meetings of the Philosophy of Science Association* 64 (Suppl.): S15–25.

Heinimann, Felix. 1945. *Nomos und Physis: Herkunft und Bedeutung einer Antithese im Griechischen Denken des 5. Jahrhunderts*. Schweizerische Beiträge Zur Altertumswissenschaft, vol. 1. Basel: F. Reinhardt.

Henrich, Joseph, Steven J. Heine, and Ara Norenzayan. 2010. The weirdest people in the world? *Behavioral and Brain Sciences* 33 (2–3): 61–83.

Hesslow, Germund. 1984. What is a genetic disease? On the relative importance of causes. In *Health, Disease, and Causal Explanations in Medicine*, edited by Lennart Nordenfelt, B. Lindahl, and B. Ingemar, 183–193. Dordrecht: Reidel.

Hesslow, Germund. 1988. The problem of causal selection. In *Contemporary Science and Natural Explanation: Common Sense Conceptions of Causality*, edited by Denis J. Hilton, 11–32. Harvester Press.

Hitchcock, Christopher, and Joshua Knobe. 2009. Cause and norm. *Journal of Philosophy* 106 (11): 587–612.

Hubbard, Ruth. 1990. *The Politics of Women's Biology*. New Brunswick, NJ: Rutgers University Press.

Hull, David L. 1973. *Darwin and His Critics: The Reception of Darwin's Theory of Evolution by the Scientific Community*. Cambridge, MA: Harvard University Press.

Hull, David L. 1978. A matter of individuality. *Philosophy of Science* 45:335–360.

Hull, David L. 1986. On human nature. *Philosophy of Science, Proceedings of the Biennial Meetings of the Philosophy of Science Association* 2 (Symposia and invited papers): 3–13.

Hume, David. (1739) 1896. *A Treatise of Human Nature*. Edited by Lewis A. Selby-Bigge. Oxford: Clarendon Press.

Hume, David (1777) 1902. *Enquiries Concerning the Human Understanding and Concerning the Principles of Morals*. 2nd ed. Edited by Lewis A. Selby-Bigge. Oxford: Clarendon Press.

Hund, Wulf D., Charles W. Mills, and Silvia Sebastiani, eds. 2016. *Simianization: Apes, Gender, Class, and Race*. Zürich: LIT Verlag.

Hursthouse, Rosalind. 2012. Human nature and Aristotelian virtue ethics. *Royal Institute of Philosophy Supplements* 70:169–188.

Inglis, David, John Bone, and Rhoda Wilkie, eds. 2005. *Nature: Critical Concepts in the Social Sciences*, vols. 1–4. London: Routledge.

Ingold, Tim. 2000. *Perception of the Environment: Essays on Livelihood, Dwelling and Skill*. New York: Routledge.

Jablonka, Eva. 2013. Some problems with genetic horoscopes. In *Genetic Explanations: Sense and Nonsense*, edited by Sheldon Krimsky and Jeremy Gruber, 71–80. Cambridge, MA: Harvard University Press.

Jablonka, Eva. 2016. Cultural epigenetics. *Sociological Review Monographs* 64 (1): 42–60.

Jablonka, Eva, and Marion J. Lamb. 1995. *Epigenetic Inheritance and Evolution: The Lamarckian Dimension*. Oxford: Oxford University Press.

Jablonka, Eva, and Marion J. Lamb. 2005. *Evolution in Four Dimensions: Genetic, Epigenetic, Behavioral and Symbolic Variation in the History of Life*. Cambridge, MA: MIT Press.

Jablonka, Eva, and Gal Raz. 2009. Transgenerational epigenetic inheritance: Prevalence, mechanisms, and implications for the study of heredity and evolution. *Quarterly Review of Biology* 84 (2): 131–176.

Jackson, John P. 2010. Definitional argument in evolutionary psychology and cultural anthropology. *Science in Context* 23 (1): 121–150.

Jaggar, Alison M. 1983. *Feminist Politics and Human Nature*. Totowa, NJ: Rowman & Allanheld.

Jaggar, Alison M., and Karsten Struhl. 2014. Human nature. In *Encyclopedia of Bioethics*, 4th ed., edited by Warren T. Reich, 1597–1612. New York: Macmillan.

Jahoda, Gustav. 1998. *Images of Savages: Ancient Roots of Modern Prejudice in Western Culture*. London: Routledge.

Jaworska, Agnieszka, and Julie Tannenbaum. 2013. The grounds of moral status. In *The Stanford Encyclopedia of Philosophy*, edited by Edward N. Zalta. Metaphysics Research Lab, Stanford University. https://plato.stanford.edu/archives/sum2013/entries/grounds-moral-status/.

Keller, Evelyn Fox. 2000. *The Century of the Gene*. Cambridge, MA: Harvard University Press.

Keller, Evelyn Fox. 2010. *The Mirage of a Space between Nature and Nurture*. Durham, NC: Duke University Press.

Kelly, Pat, and Peter Oldring. 2015. Species-less child. *This Is That*. CBC podcast, July 24. http://www.cbc.ca/radio/thisisthat/species-less-child-members-only-freeway-quiz-master-golden-opera-voice-1.3167328.

Kelman, Herbert G. 1973. Violence without moral restraint: Reflections on the dehumanization of victims and victimizers. *Journal of Social Issues* 29 (4): 25–61.

Kelsen, Hans. 1943. *Society and Nature: A Sociological Inquiry*. Chicago: University of Chicago Press.

Khalidi, Muhammad A. 2010. Interactive kinds. *British Journal for the Philosophy of Science* 61 (2): 335–360.

Khalidi, Muhammad A. 2016. Innateness as a natural cognitive kind. *Philosophical Psychology* 29 (3): 319–333.

Kitcher, Philip. 1984. Species. *Philosophy of Science* 51 (2): 308–333.

Kitcher, Philip. 1996. *The Lives to Come: The Genetic Revolution and Human Possibilities*. New York: Simon & Schuster.

Kitcher, Philip. 1999. Essence and perfection. *Ethics* 110 (1): 59–83.

Kitcher, Philip. 2001a. Battling the undead: How and (how not) to resist genetic determinism. In *Thinking About Evolution: Historical, Philosophical and Political Per-

spectives, edited by Rama Singh, Costas Krimbas, Diane B. Paul, and John Beatty, 2:396–414. Cambridge: Cambridge University Press.

Kitcher, Philip. 2001b. *Science, Truth, and Democracy*. Oxford: Oxford University Press.

Kittay, Eva F. 2008. At the margins of moral personhood. *Journal of Bioethical Inquiry* 5 (2–3): 2–3.

Kittay, Eva F., and Licia Carlson. 2010. *Cognitive Disability and Its Challenge to Moral Philosophy*. Hoboken, NJ: Wiley

Klein, Richard G. 2016. Issues in human evolution. *Proceedings of the National Academy of Sciences of the United States of America* 113 (23): 6345–6347.

Koselleck, Reinhart. (1993) 2006. Feindbegriffe. In *Begriffsgeschichten: Studien zur Semantik und Pragmatik der politischen und sozialen Sprache*, 247–284. Frankfurt am Main: Suhrkamp.

Kourany, Janet A. 2016. Should some knowledge be forbidden? The case of cognitive differences research. *Philosophy of Science* 83 (5): 779–790.

Kripke, Saul A. 1972. *Naming and Necessity*. Cambridge, MA: Harvard University Press.

Kroeber, Alfred L. 1917. The superorganic. *American Anthropologist* 19 (2): 163–213.

Kronfeldner, Maria. 2007. Is cultural evolution Lamarckian? *Biology and Philosophy* 22 (4): 493–512.

Kronfeldner, Maria. 2008. Trigger me: Evolutionspsychologie, Genzentrismus und die Idee der Kultur. *Nach Feierabend: Zürcher Jahrbuch für Wissensgeschichte* 4: 31–46.

Kronfeldner, Maria. 2009. If there is nothing beyond the organic …: Heredity and culture at the boundaries of anthropology in the work of Alfred L. Kroeber. *NTM—Zeitschrift für Geschichte der Wissenschaften, Technik und Medizin* 17 (2): 107–133.

Kronfeldner, Maria. 2011. *Darwinian Creativity and Memetics*. Durham, NC: Acumen.

Kronfeldner, Maria. 2014. How norms make causes. *International Journal of Epidemiology* 43 (6): 1707–1713.

Kronfeldner, Maria. 2015. Reconstituting phenomena. In *Recent Developments in the Philosophy of Science: EPSA13 Helsinki*, edited by U. Mäki, I. Votsis, S. Ruphy, and G. Schurz, 169–182. Cham, Switzerland: Springer.

Kronfeldner, Maria. 2017. The politics of human nature. In *On Human Nature: Evolution, Diversity, Psychology, Ethics, Politics and Religion*, edited by Michel Tibayrenc and Francisco J. Ayala, 625–632. Orlando, FL: Academic Press.

Kronfeldner, Maria. Forthcoming. Divide and conquer: The authority of nature and why we disagree about human nature. In *Why We Disagree about Human Nature*, edited by Elisabeth Hannon and Tim Lewens, 186–206. Oxford: Oxford University Press.

Kronfeldner, Maria, Neil Roughley, and Georg Toepfer. 2014. Recent work on human nature: Beyond traditional essences. *Philosophy Compass* 9 (9): 642–652.

Kropotkin, Piotr Alexeievich. 1902. *Mutual Aid: A Factor or Evolution*. New York: McClure Phillips.

Kteily, Nour, Emile Bruneau, Adam Waytz, and Sarah Cotterill. 2015. The ascent of man: Theoretical and empirical evidence for blatant dehumanization. *Journal of Personality and Social Psychology* 109 (5): 901–931.

Kupperman, Joel. 2010. *Theories of Human Nature*. Indianapolis: Hackett.

Kuzawa, Christopher W., and Elizabeth Sweet. 2009. Epigenetics and the embodiment of race: Developmental origins of US racial disparities in cardiovascular health. *American Journal of Human Biology* 21 (1): 2–15.

Laimann, Jessica. 2014. Evolutionary psychology and the nomological notion of human nature. Unpublished manuscript.

Laland, Kevin N., John Odling-Smee, William Hoppitt, and Tobias Uller. 2013. More on how and why: Cause and effect in biology revisited. *Biology and Philosophy* 28 (5): 719–745.

La Mettrie, Julien Offray de. (1748) 1912. *L'homme machine (Man a Machine)*. Edited by Gertrude Carman Bussey. La Salle: Open Court.

Lamm, Ehud. 2012. Inheritance systems. In *Stanford Encyclopedia of Philosophy*. In *The Stanford Encyclopedia of Philosophy*, edited by Edward N. Zalta. Metaphysics Research Lab, Stanford University. http://stanford.library.usyd.edu.au/entries/inheritance-systems/.

LeBon, Gustave. 1881. *L'homme et les sociétés: Leurs origines et leur histoire*. Paris: J. Rothschild.

Lennox, James G. 1987. Kinds, forms of kinds, and the more and the less in Aristotle's biology. In *Philosophical Issues in Aristotle's Biology*, edited by Allan Gotthelf, 339–359. Cambridge: Cambridge University Press.

Lerner, Henrik, and Bjørn Hofmann. 2011. Normality and naturalness: A comparison of the meanings of concepts used within veterinary medicine and human medicine. *Theoretical Medicine and Bioethics* 32 (6): 403–412.

Lévi-Strauss, Claude. 1952. *Race and History*. Paris: UNESCO.

Levins, Richard, and Richard C. Lewontin. 1985. What is human nature? In *The Dialectical Biologist*, 253–265. Cambridge, MA: Harvard University Press.

References

Lewens, Tim. 2012. Human nature: The very idea. *Philosophy and Technology* 25 (4): 459–474.

Lewens, Tim. 2015. *Cultural Evolution: Conceptual Challenges*. Oxford: Oxford University Press.

Lewis, Clive S. 1960. *Studies in Words*. Cambridge: Cambridge University Press.

Lewis, David. 1973. Causation. *Journal of Philosophy* 70 (17): 556–567.

Lewontin, Richard C. 1970. The units of selection. *Annual Review of Ecology and Systematics* 1:1–18.

Lewontin, Richard C. 1983. The organism as the subject and object of evolution. *Scientia* 77 (18): 63–82.

Lewontin, Richard C. (1974) 2006. The analysis of variance and the analysis of causes. *International Journal of Epidemiology* 35 (3): 520–525.

Linquist, S., E. Machery, P. E. Griffiths, and K. Stotz. 2011. Exploring the folkbiological conception of human nature. *Philosophical Transactions of the Royal Society of London B: Biological Sciences* 366 (1563): 444–453.

Lipton, Peter. 1990. Contrastive explanation. *Royal Institute of Philosophy* 27 (Suppl.): 247–266.

Lloyd, Geoffrey E. R. 1991. The invention of nature. In *Methods and Problems in Greek Science*, 417–434. Cambridge: Cambridge University Press.

Lloyd, Geoffrey E. R. 2011. Humanity between gods and beasts? Ontologies in question. *Journal of the Royal Anthropological Institute* 17 (4): 829–845.

Lloyd, Geoffrey E. R. 2012. *Being, Humanity, and Understanding: Studies in Ancient and Modern Societies*. Oxford: Oxford University Press.

Lock, Margaret. 2013. The epigenome and nature/nurture reunification: A challenge for anthropology. *Medical Anthropology* 32 (4): 291–308.

Longino, Helen E. 2013. *Studying Human Behavior: How Scientists Investigate Aggression and Sexuality*. Chicago: University of Chicago Press.

López-Beltrán, Carlos. 1994. Forging heredity: From metaphor to cause, a reification story. *Studies in History and Philosophy of Science* 25 (2): 211–235.

López-Beltrán, Carlos. 2007. The medical origins of heredity. In *Heredity Produced: At the Crossroads of Biology, Politics, and Culture, 1500–1870*, edited by Staffan Müller-Wille and Hans-Jörg Rheinberger, 105–132. Cambridge, MA: MIT Press.

Lorenz, Konrad. 1960. *Das sogenannte Böse*. Vienna: Dr. G. Borotha-Schoeler.

Love, Alan C. 2009. Typology reconfigured: From the metaphysics of essentialism to the epistemology of representation. *Acta Biotheoretica* 57 (1–2): 51–75.

Lovejoy, Arthur O. 1936. *The Great Chain of Being: A Study of the History of an Idea*. Cambridge, MA: Harvard University Press.

Ludwig, David. (2014). Hysteria, race, and phlogiston: A model of ontological elimination in the human sciences. *Studies in History and Philosophy of Science Part C: Studies in History and Philosophy of Biological and Biomedical Sciences* 45:68–77.

Machery, Edouard. 2008. A plea for human nature. *Philosophical Psychology* 21 (3): 321–329.

Machery, Edouard. 2012. Reconceptualizing human nature: Response to Lewens. *Philosophy and Technology* 25 (4): 475–478.

Machery, Edouard. 2016a. Human nature. In *How Biology Shapes Philosophy: New Foundations for Naturalism*, edited by David Livingstone Smith, 204–226. Cambridge: Cambridge University Press.

Machery, Edouard. 2016b. Interview on human nature. In *Conversations on Human Nature*, edited by Agustín Fuentes and Aku Visala, 55–67. London: Routledge, Taylor and Francis.

Machery, Edouard. Forthcoming. Doubling down on the nomological notion of human nature. In *Why We Disagree about Human Nature*, edited by Elisabeth Hannon and Tim Lewens, 18–39. Oxford: Oxford University Press.

Mackie, John L. 1974. *The Cement of the Universe: A Study of Causation*. Oxford: Clarendon Press.

Mameli, Matteo, and Patrick Bateson. 2006. Innateness and the sciences. *Biology and Philosophy* 21 (2): 155–188.

Matthews, Gareth B. 1986. Gender and essence in Aristotle. *Australasian Journal of Philosophy* 64 (Suppl.): 16–25.

Matthews, Gareth B. 1990. Aristotelian essentialism. *Philosophy and Phenomenological Research* 50 (October): 251–262.

Mayden, R. L. 1997. A hierarchy of species concepts: The denouement in the saga of the species problem. *Systematics Association* (special vol.) 54: 381–424.

Mayr, Ernst. 1959. Darwin and the evolutionary theory in biology. In *Evolution and Anthropology: A Centennial Appraisal*, edited by Betty J. Meggers, 1–10. Washington, DC: Anthropological Society of Washington.

Mayr, Ernst. 1961. Cause and effect in biology. *Science* 134 (3489): 1501–1506.

Mayr, Ernst. 1963. *Animal Species and Evolution*. Cambridge, MA: Harvard University Press.

Mayr, Ernst. 1976. Lamarck revisited. In *Evolution and the Diversity of Life: Selected Essays*, edited by Ernst Mayr, 222–250. Cambridge, MA: Harvard University Press.

References

McMahan, Jeff. 2002. *The Ethics of Killing: Problems at the Margins of Life*. New York: Oxford University Press.

McOuat, Gordon. 2009. The origins of "natural kinds": Keeping "essentialism" at bay in the age of reform. *Intellectual History Review* 19 (2): 211–230.

Meaney, Michael J. 2001. Maternal care, gene expression, and the transmission of individual differences in stress reactivity across generations. *Annual Review of Neuroscience* 24 (1): 1161–1192.

Meloni, Maurizio. 2015. Epigenetics for the social sciences: Justice, embodiment, and inheritance in the postgenomic age. *New Genetics and Society* 34 (2): 125–151.

Meloni, Maurizio. 2016. From boundary-work to boundary object: How biology left and re-entered the social sciences. *Sociological Review Monographs* 64 (1): 61–78.

Menzies, Peter. 2004. Difference making in context. In *Causation and Counterfactuals*, edited by John Collins, Ned Hall, and L. A. Paul, 139–180. Cambridge, MA: MIT Press.

Menzies, Peter, and Huw Price. 1993. Causation as a secondary quality. *British Journal for the Philosophy of Science* 44 (2): 187–203.

Merlin, Francesca. 2010. On Griffiths and Gray's concept of expanded and diffused inheritance. *Biological Theory* 5 (3): 206–215.

Merlin, Francesca. 2017. Limited extended inheritance. In *Challenging the Modern Synthesis: Adaptation, Development, and Inheritance*, edited by Philippe Huneman and Denis Walsh, 263–279. Oxford: Oxford University Press.

Mesoudi, Alex, Simon Blanchet, Anne Charmantier, Étienne Danchin, Laurel Fogarty, Eva Jablonka, Kevin N. Laland, et al. 2013. Is non-genetic inheritance just a proximate mechanism? A corroboration of the extended evolutionary synthesis. *Biological Theory* 7 (3): 189–195.

Mikkola, Mari. 2011. Dehumanization. In *New Waves in Ethics*, edited by Thom Brooks, 128–149. London: Palgrave-Macmillan.

Mikkola, Mari. 2016. *The Wrong of Injustice: Dehumanization and Its Role in Feminist Philosophy*. Oxford: Oxford University Press.

Milam, Erika L. Forthcoming. *Creatures of Cain: The Hunt for Human Nature in Cold War America*. Princeton, NJ: Princeton University Press.

Mill, John Stuart. 1858. *A System of Logic: Ratiocinative and Inductive*. New York: Harper & Bros.

Mill, John Stuart. 1874. Nature. In *Nature, the Utility of Religion, and Theism*, 373–402. London: Longmans, Green, Reader, and Dyer.

Millikan, Ruth G. 1999. Historical kinds and the "special sciences." *Philosophical Studies* 95 (1–2): 45–65.

Mitchell, Sandra D. 2003. *Biological Complexity and Integrative Pluralism*. Cambridge: Cambridge University Press.

Mitchell, Sandra D. 2009. *Unsimple Truths: Science, Complexity, and Policy*. Chicago: University of Chicago Press.

Müller, Andreas Th. 2015. Straßburger, Unschärferelation: Rechtsphilosophische Reflexionen zum Menschenbild von EMRK und EGMR. In *Being Human—Foundation, Imperative or Platitude? 10th Congress of the Austrian Society for Philosophy*. Innsbruck.

Müller-Wille, Staffan, and Hans-Jörg Rheinberger, eds. 2007. *Heredity Produced at the Crossroads of Biology, Politics, and Culture, 1500–1870*. Cambridge, MA: MIT Press.

Nanney, David L. 1958. Epigenetic control systems. *Proceedings of the National Academy of Sciences of the United States of America* 44 (7): 712–717.

Needham, Paul. 2008. Is water a mixture? Bridging the distinction between physical and chemical properties. *Studies in History and Philosophy of Science* 39 (1): 66–77.

Nicoglou, Antonine, and Merlin Francesca, 2017. Epigenetics: A way to bridge the gap between biological fields. *Studies in History and Philosophy of Biological and Biomedical Sciences Part C* 66: 73–82.

Norton, David F. 1993. Hume, human nature, and the foundations of morality. In *The Cambridge Companion to Hume*, edited by David F. Norton, 148–181. Cambridge: Cambridge University Press.

Nussbaum, Martha C. 1992. Human functioning and social justice: In defense of Aristotelian essentialism. *Political Theory* 20 (2): 202–246.

Nussbaum, Martha C. 1995. Aristotle on human nature and the foundations of ethics. In *World, Mind and Ethics: Essays on the Ethical Philosophy of Bernard Williams*, edited by J. E. J. Altham and R. Harrison, 61–104. Cambridge: Cambridge University Press.

Nussbaum, Martha C. 2000. Aristotle, politics, and human capabilities: A response to Antony, Arneson, Charlesworth, and Mulgan. *Ethics* 111 (1): 102–140.

Nussbaum, Martha C. 2006. *Frontiers of Justice: Disability, Nationality, Species Membership*. Cambridge, MA: Belknap Press of Harvard University Press.

Odenbaugh, Jay. 2017. Nothing in ethics makes sense except in the light of evolution? Natural goodness, normativity, and naturalism. *Synthese* 194 (4): 1031–1055.

Odling-Smee, John. 2007. Niche inheritance: A possible basis for classifying multiple inheritance systems in evolution. *Biological Theory* 2 (3): 276–289.

Odling-Smee, John. 2010. Niche inheritance. In *Evolution, the Extended Synthesis*, edited by Massimo Pigliucci and Gerd Müller, 175–207. Cambridge, MA: MIT Press.

Odling-Smee, John, and Kevin N. Laland. 2011. Ecological inheritance and cultural inheritance: What are they and how do they differ? *Biological Theory* 6 (3): 220–230.

Okasha, Samir. 2002. Darwinian metaphysics: Species and the question of essentialism. *Synthese* 131 (2): 191–213.

Olafson, Frederick A. 1995. *What Is a Human Being? A Heideggerian View*. Cambridge: Cambridge University Press.

Opotow, Susan. 1990. Moral exclusion and injustice: An introduction. *Journal of Social Issues* 46 (1): 1–20.

Oyama, Susan. 1985. *The Ontogeny of Information: Developmental Systems and Evolution*. Cambridge: Cambridge University Press.

Oyama, Susan. 2000. *Evolution's Eye: A Systems View of the Biology-Culture Divide*. Durham, NC: Duke University Press.

Özmen, Elif. 2011. Ecce homo faber! Anthropologische Utopien und das Argument von der Natur des Menschen. In *Die Gegenwart der Utopie: Zeitkritik und Denkwende*, edited by Julian Nida-Rumelin and Klaus Kufeld, 101–124. Freiburg: Alber.

Pääbo, Svante. 2001. The human genome and our view of ourselves. *Science* 291 (5507): 1219–1220.

Paul, Diane B., and Jeffrey P. Brosco. 2013. *The PKU Paradox: A Short History of a Genetic Disease*. Baltimore, MD: Johns Hopkins University Press.

Paul, Diane B., and Benjamin Day. 2008. John Stuart Mill, innate differences, and the regulation of reproduction. *Studies in History and Philosophy of Science Part C: Studies in History and Philosophy of Biological and Biomedical Sciences* 39 (2): 222–231.

Pellegrin, Pierre. 1986. *Aristotle's Classification of Animals: Biology and the Conceptual Unity of the Aristotelian Corpus*. Berkeley: University of California Press.

Pellett, Philip E., Dharam V. Ablashi, Peter F. Ambros, Henri Agut, Mary T. Caserta, Vincent Descamps, Louis Flamand, et al. 2012. Chromosomally integrated human herpesvirus 6: Questions and answers. *Reviews in Medical Virology* 22 (3): 144–155.

Pigliucci, Massimo, and Gerd Müller, eds. 2010. *Evolution, the Extended Synthesis*. Cambridge, MA: MIT Press.

Pinker, Steven. 2002. *The Blank Slate: The Modern Denial of Human Nature*. New York: Viking.

Pinker, Steven. 2011. *The Better Angels of Our Nature: Why Violence Has Declined*. New York: Viking.

Prinz, Jesse J. 2012. *Beyond Human Nature: How Culture and Experience Shape the Human Mind*. New York: Norton.

Proctor, Robert. 1995. *Cancer Wars: How Politics Shapes What We Know and Don't Know about Cancer*. New York: Basic Books.

Proctor, Robert. 2003. Three roots of human recency: Molecular anthropology, the refigured Acheulean, and the UNESCO response to Auschwitz. *Current Anthropology* 44 (2): 213–239.

Putnam, Hilary. 1975. *Mind, Language, and Reality*. Cambridge: Cambridge University Press.

Queiroz, Kevin De. 2007. Species concepts and species delimitation. *Systematic Biology* 56 (6): 879–886.

Quine, Willard V. 1969. Natural kinds. In *Ontological Relativity and Other Essays*, 114–138. New York: Columbia University Press.

Ramsey, Grant. 2013. Human nature in a post-essentialist world. *Philosophy of Science* 80 (5): 983–993.

Ramsey, Grant. Forthcoming. Trait bin and trait cluster accounts of human nature. In *Why We Disagree about Human Nature*, edited by Elisabeth Hannon and Tim Lewens, 40–57. Oxford: Oxford University Press.

Rawls, John. 1971. *A Theory of Justice*. Cambridge, MA: Belknap Press.

Reutlinger, Alexander. 2012. Getting rid of interventions. *Studies in History and Philosophy of Biology and Biomedical Science* 43 (4): 787–795.

Reutlinger, Alexander, Gerhard Schurz, and Andreas Hüttemann. 2011. Ceteris paribus laws. In *The Stanford Encyclopedia of Philosophy*, edited by Edward N. Zalta. http://plato.stanford.edu/archives/spr2011/entries/ceteris-paribus/.

Reydon, Thomas. 2009. How to fix kind membership: A problem for HPC-theory and a solution. *Philosophy of Science* 76 (5): 724–736.

Reydon, Thomas. 2012. Essentialism about kinds: An undead issue in the philosophies of physics and biology? In *Probabilities, Laws, and Structures*, edited by D. Dieks, W. J. Gonzalez, S. Hartmann, M. Stöltzner, and M. Weber, 217–230. Dordrecht: Springer Netherlands.

Reydon, Thomas. 2014. The evolution of human nature and its implications for politics: A critique. *Journal of Bioeconomics* 17 (1): 17–36.

Richardson, Sarah S., and Hallam Stevens, eds. 2015. *Postgenomics: Perspectives on Biology after the Genome*. Durham, NC: Duke University Press.

Richards, Richard A. 2010. *The Species Problem: A Philosophical Analysis*. Cambridge: Cambridge University Press.

References

Richards, Robert J. 1987. *Darwin and the Emergence of Evolutionary Theories of Mind and Behavior.* Chicago: University of Chicago Press.

Richerson, Peter J., and Robert Boyd. 2005. *Not by Genes Alone: How Culture Transformed Human Evolution.* Chicago: University of Chicago Press.

Ridley, Matt. 2003. *Nature via Nurture: Genes, Experience, and What Makes Us Human.* New York: HarperCollins.

Rieppel, Olivier. 2009. Species as a process. *Acta Biotheoretica* 57 (1–2): 33–49.

Rieppel, Olivier. 2013. Biological individuals and natural kinds. *Biological Theory* 7 (2): 162–169.

Robert, Jason S. 2008. Evo-devo. In *The Oxford Handbook of Philosophy of Biology,* edited by Michael Ruse, 291–309. Oxford: Oxford University Press.

Robertson, Teresa. 2013. Essential vs. accidental properties. In *The Stanford Encyclopedia of Philosophy,* edited by Edward N. Zalta. Metaphysics Research Lab, Stanford University. http://plato.stanford.edu/archives/fall2013/entries/essential-accidental/.

Rorty, Richard. 1999. *Philosophy and Social Hope.* New York: Penguin Books.

Roughley, Neil, ed. 2000. *Being Humans: Anthropological Universality and Particularity in Transdisciplinary Perspectives.* Berlin: Walter de Gruyter.

Roughley, Neil. 2011. Human natures. In *Human Nature and Self Design,* edited by Sebastian Schleidgen, Michael Jungert, Robert Bauer, and Verena Sandow, 11–33. Paderborn: Mentis.

Ryder, Richard D. 2010. Speciesism again: The original leaflet. *Critical Sociology* 1 (2): 1–2.

Sahlins, Marshall. 1976. *Culture and Practical Reason.* Chicago: University of Chicago Press.

Sahlins, Marshall. 2008. *The Western Illusion of Human Nature.* Chicago: Prickly Paradigm Press.

Samuels, Richard. 2012. Science and human nature. *Royal Institute of Philosophy Supplements* 70:1–28.

Sandel, Michael J. 2007. *The Case against Perfection: Ethics in the Age of Genetic Engineering.* Cambridge, MA: Harvard University Press.

Sandis, Constantine, and M. J. Cain, eds. 2012. *Human Nature.* Cambridge: Cambridge University Press.

Sarkar, Sahotra. 1999. From the Reaktionsnorm to the adaptive norm: The norm of reaction, 1909–1960. *Biology and Philosophy* 14 (2): 235–252.

Schaffner, Kenneth F. 1998. Genes, behavior, and developmental emergentism: One process, indivisible? *Philosophy of Science* 65 (2): 209–252.

Schaffner, Kenneth F. 2006. Reduction: The Cheshire cat problem and a return to roots. *Synthese* 151 (3): 377–402.

Schaffner, Kenneth F. 2016. *Behaving: What's Genetic, What's Not, and Why Should We Care?* Oxford: Oxford University Press.

Schwerin, Alexander von. 2011. From the atomic age to anti-aging: Shaping the surface between the environment and the organism. In *Membranes Surfaces Boundaries: Interstices in the History of Science, Technology and Culture*, edited by Mathias Grote and Max Stadler, 135–158. Berlin: MPIWG.

Scott-Phillips, Thomas C., Thomas E. Dickins, and Stuart A. West. 2011. Evolutionary theory and the ultimate–proximate distinction in the human behavioral sciences. *Perspectives on Psychological Science* 6 (1): 38–47.

Shapiro, Ian. 1998. Human nature. In *Routledge Encyclopedia of Philosophy*, edited by Edward Craig. London: Routledge. http://www.rep.routledge.com/article/S029.

Silvers, Anita. 1998. A fatal attraction to normalizing: Treating disabilities as deviations from "species-typical" functioning. In *Enhancing Human Traits: Ethical and Social Implications*, edited by Erik Parens, 95–123. Washington, DC: Georgetown University Press.

Simon, Herbert A. 1962. The architecture of complexity. *Proceedings of the American Philosophical Society* 106 (6): 467–482.

Singer, Peter. 1975. *Animal Liberation: A New Ethics for Our Treatment of Animals*. New York: Avon Books.

Singer, Peter. 2009. Speciesism and moral status. *Metaphilosophy* 40 (3–4): 567–581.

Smith, David Livingstone. 2011. *Less Than Human: Why We Demean, Enslave, and Exterminate Others*. New York: St. Martin's Press.

Smith, David Livingstone. 2012. War, evolution and the nature of human nature. In *The Oxford Handbook of Evolutionary Perspectives on Violence, Homicide, and War*, edited by Todd K. Shackelford and Viviana A. Weekes-Shackelford, 339–350. New York: Oxford University Press.

Smith, David Livingstone. 2013. Indexically yours: Why being human is more like being here than it is like being water. In *The Politics of Species: Reshaping Our Relationships with Other Animals*, edited by Annette Corbey and Raymond Lanjouw, 40–52. Cambridge: Cambridge University Press.

Smith, David Livingstone. 2014. Dehumanization, essentialism, and moral psychology. *Philosophy Compass* 9 (11): 814–824.

Sober, Elliott. 1980. Evolution, population thinking, and essentialism. *Philosophy of Science* 47 (3): 350–383.

Sober, Elliott. 1984. Sets, species, and evolution: Comments on Philip Kitcher's "Species." *Philosophy of Science* 51 (2): 334–341.

Sober, Elliott. (1988) 1994. Apportioning causal responsibility. In *From a Biological Point of View: Essays in Evolutionary Philosophy*, 184–200. Cambridge: Cambridge University Press.

Spencer, Herbert. 1873. *The Study of Sociology*. London: Henry King & Co.

Spencer, Herbert. 1893. The inadequacy of "natural selection," I and II. *Contemporary Review* 63 (February–March): 153–166, 439–456.

Spencer, Herbert. 1894. Weismannism once more. *Contemporary Review* 63:592–608.

Sperber, Dan. 1996. *Explaining Culture: A Naturalist Approach*. Oxford: Blackwell.

Stenmark, Mikael. 2012. Theories of human nature: Key issues. *Philosophy Compass* 7 (8): 543–558.

Stevenson, Leslie, David L. Haberman, and Peter Matthews Wright, eds. 2012. *Twelve Theories of Human Nature*. 6th ed. Oxford: Oxford University Press.

Stichweh, Rudolf. 2010. *Der Fremde: Studien zur Soziologie und Sozialgeschichte*. Berlin: Suhrkamp.

Stotz, Karola. 2010. Human nature and cognitive-developmental niche construction. *Phenomenology and the Cognitive Sciences* 9 (4): 483–501.

Stotz, Karola, and Paul Griffiths. Forthcoming. A developmental systems account of human nature. In *Why We Disagree about Human Nature*, edited by Elisabeth Hannon and Tim Lewens, 58–75. Oxford: Oxford University Press.

Strathern, Marilyn. 1992. *After Nature: English Kinship in the Late Twentieth Century*. Cambridge: Cambridge University Press.

Sumner, William G. 1906. *Folkways: A Study of the Sociological Importance of Usage, Manners, Customs, Mores and Morals*. Boston: Ginn.

Tabery, James. 2014. *Beyond versus: The Struggle to Understand the Interaction of Nature and Nurture*. Cambridge, MA: MIT Press.

Thompson, Michael. 2008. *Life and Action: Elementary Structures of Practice and Practical Thought*. Cambridge, MA: Harvard University Press.

Tomasello, Michael. 2008. *Origins of Human Communication*. Cambridge, MA: MIT Press.

Tooby, John, and Leda Cosmides. 1992. The psychological foundations of culture. In *The Adapted Mind: Evolutionary Psychology and the Generation of Culture*, edited by

Jerome H. Barkow, Leda Cosmides, and John Tooby, 19–136. Oxford: Oxford University Press.

Trigg, Roger. 1999. *Ideas of Human Nature: An Historical Introduction*. Oxford: Blackwell.

Van Fraassen, Bas C. 1980. *The Scientific Image*. Oxford: Oxford University Press.

Viveiros de Castro, Eduardo. 1992. *From the Enemy's Point of View: Humanity and Divinity in an Amazonian Society*. Chicago: University of Chicago Press.

Viveiros de Castro, Eduardo. 1998. Cosmological deixis and Amerindian perspectivism. *Journal of the Royal Anthropological Institute* 4 (3): 469–488.

Walsh, Denis. 2006. Evolutionary essentialism. *British Journal for the Philosophy of Science* 57 (2): 425–448.

Wasserman, David, Adrienne Asch, Jeffrey Blustein, and Daniel Putnam. 2013. Cognitive disability and moral status. In *The Stanford Encyclopedia of Philosophy*, edited by Edward N. Zalta. Metaphysics Research Lab, Stanford University. http://plato.stanford.edu/archives/fall2013/entries/cognitive-disability/.

Waters, C. Kenneth. 2000. Molecules made biological. *Revue Internationale de Philosophie* 214 (4): 539–564.

Waters, C. Kenneth. 2007. Causes that make a difference. *Journal of Philosophy* 104 (11): 551–579.

Weber, Marcel. Forthcoming. Causal selection vs causal parity in biology: Relevant counterfactuals and biologically normal interventions. In *Philosophical Perspectives on Causal Reasoning in Biology*, edited by C. Kenneth Waters and James Woodward. Minnesota Studies in Philosophy of Science, vol. 21. Minneapolis: University of Minnesota Press.

Weismann, August. 1891. *Essays upon Heredity and Kindred Biological Problems*. Oxford: Clarendon Press.

Weismann, August. 1893. The all-sufficiency of natural selection: A reply to Herbert Spencer. *Contemporary Review* 64: 309–338, 596–610.

Wells, Jonathan C. K. 2010. Maternal capital and the metabolic ghetto: An evolutionary perspective on the transgenerational basis of health inequalities. *American Journal of Human Biology* 22 (1): 1–17.

West-Eberhard, Mary J. 2003. *Developmental Plasticity and Evolution*. Oxford: Oxford University Press.

Wilkins, John S. 2013. Essentialism in biology. In *The Philosophy of Biology*, edited by Kostas Kampourakis, 395–419. History, Philosophy and Theory of the Life Sciences 1. Springer Netherlands.

Williams, Bernard A. O. 2006. The human prejudice. In *Philosophy as a Humanistic Discipline*, edited by A. W. Moore, 135–152. Princeton, NJ: Princeton University Press.

Williams, Raymond. (1976) 2011. *Keywords: A Vocabulary of Culture and Society*. London: Routledge.

Wilson, Robert A. 2018. *The Eugenic Mind Project*. Cambridge, MA: MIT Press.

Wilson, Robert A., Matthew J. Barker, and Ingo Brigandt. 2007. When traditional essentialism fails: Biological natural kinds. *Philosophical Topics* 35 (1/2): 189–216.

Wilson, Robert A., and Joshua St. Pierre. 2016. Eugenics and disability. In *Rethinking Disability: World Perspectives in Culture and Society*, edited by Patrick Devlieger, Beatriz Miranda-Galarza, Steven E. Brown, and Megan Strickfaden, 93–112. Antwerp-Apeldoorn: Garant Publishing.

Wimsatt, William, and James R. Griesemer. 2007. Reproducing entrenchments to scaffold culture: The central role of development in cultural evolution. In *Integrating Evolution and Development: From Theory to Practice*, edited by Roger Sansom and Robert N. Brandon, 227–323. Cambridge, MA: MIT Press.

Winsor, Mary P. 2006. The creation of the essentialism story: An exercise in metahistory. *History and Philosophy of the Life Sciences* 28 (2): 149–174.

Witt, Charlotte. 2005. Form, normativity and gender in Aristotle: A feminist perspective. In *Feminist Reflections on the History of Philosophy*, edited by Lilli Alanen and Charlotte Witt, 117–136. Dordrecht: Springer.

Wolbring, Gregor. 2008. The politics of ableism. *Development* 51 (2): 252–258.

Wood, Bernard, and Eve Boyle. 2017. Hominins: Context, origins, and taxic diversity. In *On Human Nature: Evolution, Diversity, Psychology, Ethics, Politics and Religion*, ed. Michel Tibayrenc and Francisco J. Ayala, 17–44. Orlando, FL: Academic Press.

Woodward, James. 2003. *Making Things Happen: A Theory of Causal Explanation*. New York: Oxford University Press.

Zirkle, Conway. 1946. The early history of the idea of the inheritance of acquired characters and of pangenesis. *Transactions of the American Philosophical Society* 35 (2): 91–151.

Index

Abstraction
  averaging out of culture, 160–163
  averaging out of nature, 158–160
  and capacities, 176–178
  and the disjunctive regress argument, 133–134, 137–139
  and generalizations about humans (typicality of traits), 132–134, 137, 160–166, 224–225
  pragmatic stopping of, 137–139, 162–164
Accounts of human nature
  all-inclusive, 133–136, 138–139, 164, 181–184
  dialectic, 225–226, 228, 240, 242
  interactive, 8, 9, 85–88, 148–149, 154–157, 160–163, 172–174, 200–201, 206–208, 210–211 (*see also* Interactionist consensus)
  monist vs. pluralist, 5–6, 7–10, 93, 96, 114–119, 126–129, 130, 141–145, 177–178, 181–182, 184, 187, 210–212, 233
  post-essentialist, 9–10, 95–96, 122, 126–127, 167, 175, 179–180, 184–187, 210–211, 233–238, 242 (*see also* Essentialism)
Ambiguity of terms, 115–116, 226–227, 229, 238–239
Amundson, Ron, 166, 220–221, 248n2
André, Jean-Baptiste, 107–109

Animal–human divide, 8, 29, 131–132, 142–143, 211
Animal rights, 262n4
Antony, Louise M., 31, 117, 128, 137, 223–224
Apportioning causal responsibility. *See under* Causation
Aristotle, 19–21, 27, 29, 49–50, 54, 129, 175, 177–178, 248n2
Aspects of human nature concepts, 3–4, 11, 165–166
  and family resemblance between human nature concepts, 4
  fixity (*see* Fixity of traits; Stability)
  normalcy (*see* Normalcy of traits)
  species specificity (*see* Species specificity of traits)
  typicality (*see* Typicality of traits)
Ayala, Francisco J., 124

Barker, Matthew J., 94
Bateson, Patrick, 148
Beauvoir, Simone de, 135
Biological inheritance. *See* Channels of inheritance
Black-ape facilitation effect, 22–24, 27, 29
Blank slate metaphor, 185–186
Body height, 51, 70–76, 79, 81, 158–160, 193, 209
Boulter, Stephen J., 179–180

Boyd, Richard, 95, 145
Brigandt, Ingo, 94
Bruno, Giordano, 21

Cancer, 202–204, 208
Capacities of individuals
  and abstraction, 176–178
  and dehumanization, 50
  and essentialist discounting of
    variation, 40, 51–57, 176–178
  vs. population-level kind propensities,
    54–57, 178–180
Caspi, Avshalom, 78, 81
Cassirer, Ernst, 226, 228
Causal selection
  conservative control vs. forward
    control in, 199–200
  and disagreement regarding causal
    explanations, 201–202
  and looping effects on human nature,
    204, 206–210
  normality vs. controllability account
    of, 191–195, 196–199
  and normative aspects, 201–202,
    208–210
  and partial explanation, 190–191
  prioritization of biological factors in,
    200–201, 202–206
  willingness to control, 199–202
Causation
  actual vs. potential difference making,
    194–198
  apportioning causal responsibility,
    62–64
  causal selection, 192–198
  as difference making vs. production,
    70–71, 73–75, 158–161, 164–165,
    185, 211
  (in-)commensurability of causal
    contribution, 62–64, 72–74
  nature and nurture as separate causes,
    59–64
  parity of, 53–54, 58, 87

Cavalli-Sforza, Luigi L., 106
Channels of inheritance (biological and
    cultural inheritance)
  and autonomy of culture, 64–67,
    104–106
  in coevolution, 82–85
  content (in-)dependence of
    transmission in, 106–110, 157–158
  as distinct, 103–104, 108, 113,
    156–157, 165
  and epigenetic inheritance,
    111–112
  evolutionary scope of, 110–111
  and genealogy, 102–104
  mode change in, 107–108
  and nature–nurture (nature–culture)
    distinction, 84–85, 102–103, 140,
    157, 158–163, 185, 211
  near-decomposability of, 104–106,
    112–113, 157, 186
  and niche inheritance, 112
  stability of, 8, 99, 108–112, 114–115,
    123–124, 141, 144–145, 149–150,
    154–158, 166, 177
  vertical, horizontal, and oblique mode
    in, 106–109
Claidière, Nicolas, 107–109
Classificatory human nature, 4–8
  vs. descriptive nature, 114–116,
    126–129
  explanatory nature, its relation to,
    113–116, 187
  genealogical nexus as, 4, 7, 36–37,
    41, 43–44, 47–49, 91, 93–102, 114,
    127–128
  and the problem of squaring the
    circles, 45–49
  and questions related to species (see
    under Species)
  and new reproductive technologies
    (see under Reproduction)
  and the tautology problem, 47–48,
    101

Index 291

and the transitivity argument,
    100–101
Coevolution of nature and culture,
    82–85
  and evolvedness of human nature
    traits, 153–157
  and externalization of vitamin C
    production, 173
  and lactose (in-)tolerance, 84, 105
  and near-decomposability, 104–106
Collingwood, Robert G., 189–192,
    197–200, 202
Content of human nature concept, 26
  deliberation about importance of,
    221–224, 228
  disagreement on, 140–144
  (ex-)changeability of, 15, 21, 28–31,
    237
  objectivity of, 15, 32, 87
  and property-based dehumanization,
    26–28
Contrastive power, 134, 138–139,
    181–182, 183, 186, 212. *See also*
    Predictive power
Creating human nature. *See* Looping
    effects
Cultural evolution
  and anti-Lamarckism, 64–67
  as part of coevolution (*see*
    Coevolution of nature and culture)
Cultural inheritance. *See* Channels of
    inheritance
Culture
  autonomy from nature, 64–67,
    104–106
  cultural determinism, 65, 84
  as a fast track of evolution, 66–67,
    108–114

Danchin, Étienne, 109, 112
Darwin, Charles, 22, 61–62, 64, 132
Darwinism. *See also* Population
    thinking; Essentialism

  and racism, 22
  and variational vs. transformational
    evolution, 38–40
Daston, Lorraine, 245n9
Day, Benjamin, 63–64
Dehumanization, 16–19, 28–29, 31–32
  animalistic and mechanistic, 30–31
  of early hominids, 27–28, 97–98
  and elimination of the term *human
    nature* (*see under* Elimination of the
    term *human nature*)
  and essentialism, 20, 58, 234–237
  as a graded spectrum, 19, 23
  implicit and explicit, 23, 26
  of non-Europeans, 21–24, 29, 236 (*see
    also* Black-ape facilitation effect)
  reciprocity of, 30–31
  relational and property-based, 26–28,
    219, 234–236
  and the vernacular concept of human
    nature, 16–18
  of women (sexism), 18, 19–21, 25, 29,
    50, 241
Dennett, Daniel C., 176
Descriptive human nature, 8–10. *See
  also* Content of human nature
    concept
  as a cluster of generalizations about
    humans, 121–126, 127–129, 145,
    164, 234–235, 237–238
  and dehumanization, 234–235,
    237–238
  ethical importance of, 220–225
  as an etiological category, 129–131,
    167
  and the four aspects, 165–166
  humanness as, 8–9, 41, 48
  as indispensable for sciences, 124–125
  and looping effects (*see* Looping
    effects)
  as methodological issues regarding
    generalizations about humans, 123,
    141 (*see also* Abstraction)

Descriptive human nature (cont.)
  and population thinking, 126
  and the reference class, 127–129
  its relation to classificatory human nature, 114–116, 126–129
  its relation to explanatory nature, 114–116, 129–131, 173, 175
  and stability, 123–124, 139–141, 144–145, 154–185, 164–165
  thickness of, 86–87, 136, 137, 149
  and typicality (see Typicality of traits)
Desmond, Adrian J., 22
Development
  and the brick-mortar analogy, 72–74
  developmental fixity (innateness), 148–149
  developmental interaction, 70–79 (see also Causation; Gene–environment interaction; Interactionist consensus)
  developmental plasticity, 68, 149, 166, 174–175
  developmental systems theory, 59, 68–69, 84–87, 113, 151, 181–182
  and nature vs. nurture (see Gene–environment interaction)
  as a necessary part of transformational evolution, 40
  as part of intrinsic essentialism, 171–174, 175
  as part of ultimate explanation, 150–151
  partitioning of developmental resources, 62–64, 84–85, 113 (see also Channels of inheritance)
Devitt, Michael, 57, 93, 100, 169, 170–173, 175, 177–179
Dialectical biology, 68
Differences. See also Racism; Similarity; Variation
  biological, 66–67, 69, 70–71, 74–76, 158–163, 204, 206
  cultural, 65–67, 69, 70–71, 74–76, 125, 158–163

patterns of similarity and difference, 134, 139, 181–182, 206–208
Disability, 19, 41–42, 58, 217, 220–221
Disagreement
  about causal explanation, 74–76, 197–198
  about human nature concept (see Essential contestedness of the human nature concept)
  about importance of traits for human nature, 140–144, 221–224, 228
Disciplinary perspectives, 74–76, 94–95, 97–98, 125, 132, 140–144, 187
Disjunctive regress argument. See under Abstraction
Dobzhansky, Theodosius G., 142, 155
Down syndrome, 205–207
Dualistic connotations of term nature, 2–3. See also Nature–nurture distinction
Dumsday, Travis, 94, 117
Dupré, John, 98, 125, 254n4, 255n7

Elimination of the term *human nature*
  and dehumanization, 234–237
  and impossibility of neutralization, 234–237
  and the precautionary principle, 240–241
  as a regulative ideal, 232–233
  vs. revision, 232, 238–241
  and the trade-offs involved, 238–240
  as value-laden, 238–241
Embodiment. See under Epigenetics
Enhancement. See Human enhancement
Epigenetics
  and biological reproduction, 80–82, 111–112
  and diminishing importance of genes, 69, 80–82
  and embodiment, 80–82, 111
  epigenetic inheritance, 111–12

and nature–nurture distinction, 68–69, 79–82
and parental effects, 81–82, 111
and racial inequality, 204
Epistemic roles of human nature concepts, 9–10, 34, 36, 96, 122, 187, 244n7
  classificatory role, 41, 48, 126–127 (*see also* Classificatory human nature)
  descriptive and predictive role, 55, 126–127 (*see also* Descriptive human nature)
  explanatory role, 49–50 (*see also* Explanatory human nature)
  and the normalcy aspect, 16–17
Epistemic values
  clarity, 228, 235, 238–239
  parsimony, 10, 48, 178, 228
  unity, 238–239
Equality-establishing function of the human nature concept, 239–241
Ereshefsky, Marc, 34
Escherichia coli (E. coli), 117, 119
Essences. *See* Essentialism
Essential contestedness of the human nature concept, 215, 225–229
Essentialism, 3, 9–10
  in Aristotle, 19–21, 50, 175, 248n2
  and capacities as discounting variation (*see under* Capacities of individuals)
  classificatory, 9, 35–40, 41–49, 55, 95–96
  and dehumanization, 20, 58, 234–237
  descriptive-predictive, 9, 34, 38, 55
  and dispensability of essences, 52–53
  explanatory, 9, 40, 49–58, 170–179
  as incompatible with Darwinian theory, 53–54, 56–57
  intrinsicality bias in, 170–178
  metaphysical, 57
  natural state model, 49–53
  neo-essentialism, 170–178
  normative, 220–221

  psychological, 57, 247n21
  relational, 95–96
  teleological, 20, 50–51, 174–178
Ethical intrinsicalism vs. externalism. *See* Humanism
Ethnocentrism 29–30. *See also* Dehumanization
Evidential indicators. *See* Logical vs. evidential criteria
Evolution
  cultural (*see* Cultural evolution)
  of humans, 97–98
  of nature and culture (*see* Coevolution of nature and culture)
  variational vs. transformational, 38–40
Evolutionary psychology, 138, 164, 184
Evolvedness of traits
  as a successor for innateness, 150–156
  and the vagueness problem, 153–154
Explanation
  contrastive, 152, 162
  developmental 92–3, 116, 151 (*see also* Development)
  evolutionary, 82–84, 92, 130, 150–152, 155
  explanatory asymmetry regarding channels, 110
  indirect, 93–94, 116
  partial, 93–94, 116, 190–191
  ultimate vs. proximate, 150–152, 184
Explanatory human nature, 1, 8–9, 169
  and classificatory human nature, 187
  as a cluster of biologically inherited developmental resources, 94, 114, 169, 180, 184–186 (*see also* Channels of inheritance)
  and descriptive human nature, 114–116, 129–131, 173, 175, 187
  and the developmentalist challenge, 85–87
  and the "due to nature" talk, 59, 86

Explanatory human nature (cont.)
  and intrinsicality bias, 86–87, 170–178
  and partial explanation (*see under* Causal selection)
  as a property of a population, 8, 179–186
Explanatory power, 110, 130
Extrapolation. *See under* Gene–environment interaction

Feldman, Marcus W., 106
Fish, Eric W., 81
Fixity of traits, 3, 8, 10
  developmental (innateness), 148–149
  evolutionary, 150, 166
  as incompatible with interactionism, 8, 149
  vs. stability, 8, 144–145, 147, 150
Foot, Philippa, 220
Fukuyama, Francis, 75, 148, 158, 226

Gallie, Walter B., 226–228
Galton, Francis, 33, 52, 61–64, 73, 180, 232
  anti-Lamarckism, 62–64
  and the hardening of the nature–nurture divide, 61–64
  and particulate inheritance, 62
  and population thinking, 33, 52, 61–62, 180
Gannett, Lisa, 75, 193, 196, 198, 205
Genealogical nexus. *See* Classificatory human nature
Genealogy
  and channels of inheritance, 102–104
  classificatory role of (*see* Classificatory human nature)
  and dehumanization, 26, 235
  explanatory role of, 99, 116, 178
  vs. similarity, 43–45, 95, 218–220
  social and moral importance of, 45, 98, 117–119, 218–220

Gene–environment interaction. *See also* Causation; Interactionist consensus; Nature–nurture distinction;
  epigenetic (*see under* Epigenetics)
  narrow vs. broad (trivial), 77–78
  and patterns of (in-)dependence, 74–78, 162–163, 181
  and the problem of extrapolation, 76, 78–79, 161–162
  as trivial (genetic inertness), 71
Generalizations about humans. *See under* Abstraction
Generics, 16–17, 176, 245n5
Genes
  diminishing importance of, 69, 80–82
  as the essentialist explanatory nature, 56, 59, 69
  expression of 68–69, 70–71, 74–79
  genetic determinism, 59, 69, 76, 86
  interaction with environment (*see* Gene–environment interaction)
  and molecularization of life, 200, 204–206, 261n12
  pluralism in the gene concept, 227
Goff, Phillip Attiba, 22–24, 27, 247n18
Gray, Russell D., 84, 113, 159–160
Griffiths, Paul E., 3, 46, 59, 84–85, 95, 113, 124, 139–140, 145, 148, 159–160, 181–182

Hacking, Ian, 35, 189, 206
Hart, Herbert L. A., 190, 192, 194–195, 198, 260n6
Haslam, Nick, 19, 30
Heine, Steven J., 123
Henrich, Joseph, 123
Heredity. *See also* Inheritance
  emergence of the concept, 61–62
Heritability
  in the broad evolutionary sense, 83–84
  in the narrow statistical sense, 63, 73, 252n11

Index

Hippocratics, 60
Historicity (spatiotemporal restrictedness) of kinds. *See under* Kinds
Homeostasis. *See also* Stability
  homeostatic mechanisms, 95, 99, 102, 145, 183–184
  homeostatic property clusters, 95, 145
Homo sapiens (*H. sapiens*)
  conditions for membership in, 4–5, 6–7, 34, 41–47, 100–102, 216–219
  origins of (*see* Human origins)
  and the problem of squaring the circles, 45–49, 179–180
Homunculi, 21, 29, 246n16
Honoré, Tony, 190, 192, 194–195, 198, 260n6
Hull, David L., 4–5, 16, 32, 34, 36–37, 44, 47–49, 53–54, 91, 95, 100–101, 121–123, 128, 133, 141, 176, 178, 185, 212
Human
  as a conceptual blank mold, 31
  vs. person, 4–5, 7, 27, 42–45, 222, 243n4
  as pointing to a biological or social group, 4–7, 31–32, 216–217
Human enhancement, 17, 118, 208, 235, 249n9
Human fetuses, 42–43
Human flourishing, 17, 19, 220, 222, 225
Human hand, 124–125
Humanism
  internalist, 221–223
  veil-of-ignorance, 222–223
Humanity, 5–8. *See also* Classificatory human nature
  and moral standing, 7, 216–220
  overlap with humankind, 6, 216–217
Humankind, 4–8, 10. *See also* Classificatory human nature

  conditions for membership in (*see under Homo sapiens*)
  and moral standing, 216–220
  overlap with humanity, 6, 216–217
Human nature talk. *See* Elimination of the term *human nature*
Humanness. *See under* Descriptive human nature
Humanoids
  in *Blade Runner* (film), 7
  and humanness as insufficient for membership in *H. sapiens*, 41, 43–44, 48–49, 102
  and moral standing 7, 218, 219
Human origins, 97–98
Human rights, 7, 17, 32, 100, 217–220, 255n6
Hume, David, 2, 29, 115, 124

Identity
  numerical, 44, 128–129
  qualitative, 129
Incommensurability of nature and nurture. *See under* Causation
Individuality, 39–40, 170, 178. *See also* Population thinking; Variation
Inheritance. *See also* Heredity
  of acquired characteristics, 60–66
  biological and cultural subsystems of (*see* Channels of inheritance)
  epigenetic, 79–82, 111–112
  hard vs. soft, 64, 82
  niche, 112
  particulate, 62
Innateness, 148–149. *See also* Fixity of traits
Interactionist consensus, 8, 70–71. *See also* Causation; Development; Gene–environment interaction; Nature–nurture distinction
  epigenetic interaction and inheritance (*see under* Epigenetics)

Interactionist consensus (cont.)
  and genetic inertness, 71
  and (in-)commensurability of nature and nurture, 62–64, 72–74
  and interaction at the evolutionary level (see Coevolution of nature and culture)
  and quantifiable difference making, 73–79
Intrinsicality bias. See under Essentialism

Jablonka, Eva, 69, 80, 110, 111–112

Keller, Evelyn Fox, 59, 62–64, 70, 75, 185
Kelsen, Hans, 216
Kinds
  etiological, 130, 140, 167
  etymology of the term *kind*, 118
  historicity (spatiotemporal restrictedness) of, 36–38
  natural (see Essentialism; Natural kinds)
Kitcher, Philip, 76, 98, 198, 205
Klein, Richard G., 244n6
Koselleck, Reinhart, 31
Kroeber, Alfred L., 65–68, 84, 110, 158

Lactose (in-) tolerance, 84, 105, 153–154, 196
Laland, Kevin N., 112, 151, 155
Lamarck, Jean-Baptiste, 38, 40
Lamarckism
  and cultural evolution, 64–67
  and epigenetics, 69, 80
  and inheritance of acquired characteristics, 60, 61–64
  opposition to, 61–67
  as supporting racism, 66
  and transformational vs. variational evolution, 38–40
Lamb, Marion J., 69, 110, 112
La Mettrie, Julien Offray de, 46

Language, 45–46, 137, 143–144, 149, 160–163, 166, 176
LeBon, Gustave, 246n14
Lennox, James G., 49, 248n2
Levins, Richard, 181
Lévi-Strauss, Claude, 30
Lewens, Tim, 133–134, 138–139, 156–157
Lewis, Clive S., 118
Lewis, David, 191
Lewontin, Richard C., 39, 68, 72, 83, 181, 252n11
Life form of a species. See Descriptive human nature
Lloyd, Geoffrey E. R., 30
Lock, Margaret, 79–80
Logical vs. evidential criteria, 49, 95–97, 98, 101, 128
Looping effects, 9–10, 189–190, 204, 206–210, 223–224
Lopez-Beltran, Carlos, 61

Machery, Edouard, 121, 124, 129–131, 133–135, 137–140, 142, 150–153, 155–158, 162–165, 175, 182–183, 244n7
Malaria, 191, 193, 195, 207
Mameli, Matteo, 148
Mayr, Ernst, 34, 38, 49, 150
Meloni, Maurizio, 79, 80
Merlin, Francesca, 81, 104, 110
Milam, Erika L., 142–143
Mill, John Stuart, 63, 73–74, 148, 191, 260n6
Millikan, Ruth G., 95
Molecularization of life. See under Genes
Monoamine oxidase A (MAOA) gene, 78, 81
Moore, James R., 22, 235
Moral standing, 7, 216–220
  the double-entry solution, 217–220
  and speciesism, 216, 218, 219
Müller-Wille, Staffan, 61–62

Index

Naturalism vs. antinaturalism, 2–3, 249n9
Natural kinds, 34–35. *See also* Kinds
  vs. biological species, 34–37
  chemical kinds as, 35–39
  plurality of concepts of, 35
Natural state model. *See under* Essentialism
Natural vs. unnatural, 245n9
Nature
  contrastive character of the concept, 2–3
  history of the term, 1–2
  multiplicity of meanings, 1–2
  as not making a difference (averaging out) (*see under* Abstraction)
  as universal vs. racism, 66, 204, 208
Nature–nurture distinction
  and alternative ontologies of nature and culture, 67–68
  and anti-Lamarckism, 64–67
  and channels of inheritance (*see under* Channels of inheritance)
  and concept of human nature, 85–87
  and development (*see* Gene–environment interaction)
  and epigenetics, 68–69, 79–82
  as introduced by Galton, 61–64, 73
  hardening of the divide, 61–64
  nature and nurture as separate causes, 59–64 (*see also* Causation)
  physis vs. *nomos*, 60–61
Near-decomposability of systems, 104–106, 112–113, 157, 186
Necessary and sufficient conditions of membership. *See also* Moral standing
  in humanity (personhood), 5
  in humankind (*H. sapiens*), 4, 34, 40, 41–49
Neo-essentialism. *See under* Essentialism
Niche, 112
  its construction, 68, 83, 99, 112, 151, 173, 184
  its inheritance, 112
  and persistence of ecological factors, 83, 102–104
Nonlinearity. *See under* Norms of reaction
Norenzayan, Ara 123
Normalcy of traits, 3–4, 10–11, 211
  and causal explanation, 193, 195–196, 201, 203
  created via disciplining and homogenizing people, 207–208
  and dehumanization, 16–18, 26–27, 58, 237
  and normative essentialism, 16–18, 220–221
  as statistical, 51–57, 166
Normative essentialism. *See under* Essentialism
Normativity. *See also* Normalcy of traits
  as absent from the non-essentialist human nature, 166
  as applying to humans and not other species, 16–18
  and causal selection (*see under* Causal selection)
  normative human nature, 10, 220–225, 225–229
Norms of reaction. *See also* Gene–environment interaction
  for body height, 70–71, 74–76
  and difference making, 70–71, 74–79
  flat, 76, 79, 86, 135
  nonlinear, 76–77
  and patterns of (in-)dependence, 74–78, 162–163, 181
Nussbaum, Martha C., 17, 221–224, 239

Objectivity
  despite disagreement and deliberation, 141, 223–4
  of the post-essentialist natures, 89–90, 100

Objectivity (cont.)
  of the scientific concept vs. social perspectivity of the vernacular concept, 15, 32, 87
Odenbaugh, Jay, 220–221
Odling-Smee, John, 112
Okasha, Samir, 44, 47, 95–96, 102
Ontological status (reality) of human nature(s), 2–3, 9, 89–90, 118, 184–185
  and abstraction, 176–178
  and channels of inheritance, 103, 104–106
  as resulting from explanatory looping effects, 206–210, 223–224
Oyama, Susan, 59, 68

Paracelsus, 21
Paul, Diane B., 63–64
Person. *See under* Human
Phenylketonuria (PKU), 76, 134–135, 258n9
Pluralism. *See also* Disagreement
  about developmental resources, 83, 187 (*see also* Interactionist consensus)
  about epistemic values, 238–239
  vs. monism regarding concepts of human (*see under* Accounts of human nature)
  and pragmatic aspects regarding polymorphism, 138–139
  about pragmatic functions, 239–240
  about reference class, 5–6, 127–129, 210, 218
  of species concepts, 35, 98–99
Polygenism. *See under* Racism
Polymorphisms. *See under* Typicality of traits
Population thinking, 33, 38–40, 49–57, 63, 126–127, 176–179. *See also* Essentialism; Individuality; Typological thinking
  in explanatory contexts 49–57 (*see also* Explanatory nature)
  in Galton, 33, 52, 61–62, 180
  and variational vs. transformational evolution, 38–40
Pragmatic functions of human nature concepts
  and dehumanization, 15–17, 28–32
  and establishing equality, 239–241
  and human flourishing, 220–225
  and moral standing, 217–220
  and normalcy aspect, 10, 16–18, 220–221
  and styles of inquiry, 187
  trade-offs regarding, 238–242
Predictive power, 134–136, 138–139
Proctor, Robert, 16, 28, 97–98, 202–203
Property cluster. *See* Descriptive human nature; Homeostasis

Queiroz, Kevin De, 96, 98

Racism
  and dehumanization of non-Europeans, 21–24, 29, 236 (*see also* Black-ape facilitation effect)
  and polygenism, 22
  as a result of embodiment, 80
  as a result of looping effects, 204, 208
  vs. unity of humankind, 66
Ramsey, Grant, 133–136, 138–140, 142, 149
Rationality, 4, 20, 26, 40–41, 50, 86, 141, 143, 217, 249n9
Rawls, John, 5, 222–223
Reconstituting phenomena, 75, 160–164, 211
Reference class, 55–56, 127–129, 193–196
Reproduction (biological)
  and channels of inheritance, 61, 103–106

Index   299

and dehumanization in Aristotle, 20
and epigenetics, 80–82, 111–112
and genealogy, 91, 114
and nature, concept of, 2
new reproductive technologies,
    117–118
Rheinberger, Hans-Jörg, 61–62
Richards, Richard A., 98
Richards, Robert J., 248n2
Rieppel, Olivier, 35, 44, 95
Rorty, Richard, 226, 228
Roughley, Neil, 3, 20, 48, 143–144,
    244n7

Sahlins, Marshall, 30–31, 68
Samuels, Richard, 44, 130–131, 140,
    182–184, 244n7
Sartre, Jean-Paul, 135
Schaffner, Kenneth F., 59, 68, 252n11
Schwerin, Alexander von, 202–203
Sexism. *See under* Dehumanization
Shared intentionality, 144
Similarity, 38, 42, 118, 163, 180–182,
    218–219. *See also* Abstraction
and genealogy, 26–27, 44–45, 95–96,
    102, 117–119
vs. identity, 38
and inheritance, 84
moral relevance of, 218–219
patterns of similarity and difference,
    134, 139, 181–182, 206–208
Simon, Herbert A., 105
Smith, David Livingstone, 31, 236,
    246n11, 263n5
Sober, Elliott, 34, 36–37, 42, 49, 53, 56,
    63, 72, 79, 94, 129, 172, 176–179,
    252n11
Social inclusion/exclusion. *See*
    Dehumanization
Social perspectivity. *See under* Vernacular
    concept of human nature
Species
anti-essentialism about, 34–41

and the constitution question, 92–93,
    99
as defined by genealogical
    relationships (historicity), 36–38,
    93–99
and the description question, 92–93,
    116
ecological concept of, 98
as individuals, 35
necessary and sufficient conditions for
    membership in, 34, 40, 41–49
and the partaking question, 92–93,
    100–102, 171 (*see also* Classificatory
    human nature)
and plurality of species concepts,
    34–35, 98
speciation events, 96–98
and the species question, 92–93,
    96–99
and the trait explanation question,
    92–93, 116, 171–172
Speciesism
and essentialism, 235
and moral standing, 216, 218, 219
Species-less child, 119, 125, 166–167
Species specificity (uniqueness) of traits,
    3–4, 8. *See also* Animal-human
    divide
and classification, 38, 41, 45, 54
as dispensable for descriptive human
    nature, 165
as dispensable for explanatory nature,
    186
epistemic interest in, 142–142
as exclusive typicality, 131–132
Spencer, Herbert, 21, 29, 64, 66
Stability, 8
of biological inheritance, 8, 99,
    108–112, 114–115, 123–124, 141,
    144–145, 149–150, 154–158, 166,
    177
and evolutionary scope, 110–111,
    113–114

Stability (cont.)
  evolutionary vs. developmental, 149–150
  vs. fixity, 8, 144–145, 147, 150
  and homeostasis, 99 (*see also* Homeostasis)
  and nature–culture distinction, 164–165
  and typicality, 123–124, 139–141, 144–145
Stotz, Karola, 139–140, 181–182
Sumner, William G., 29
Superchimp, 46
Swamp man, 44, 48

Taxonomic relationalism, 95–96
Thompson, Michael, 220
Toepfer, Georg, 144, 244n7
Tomasello, Michael, 144
Tool use, 27–28, 124
Transhumanism. *See* Human enhancement
Transitivity argument, 100–101
Typicality of traits, 3, 8, 18, 38, 41
  and abstraction (*see under* Abstraction)
  diachronic dimension of (*see* Stability)
  ethical importance of, 220–225
  as (in)sufficient for descriptive human nature, 139–145
  as necessary for descriptive human nature, 131–139
  and polymorphic traits, 125, 132–134, 137–138
  and the problem of squaring the circles, 45–49, 179–180
  vs. species specificity, 131–132
  and stability, 123–124, 139, 144–145
  vs. universality, 126, 166, 185
Typological thinking. *See also* Essentialism
  as discounting of variation, 40, 51–57
  and transformational evolution, 38–40

  as typological individualism, 170–171, 178

Ultimate vs. proximate explanations. *See under* Explanation
Underdetermination
  and the choice of important human properties, 141
  of claims about origins of species, 97
  of culture by nature, 65–66
Uniqueness of human traits. *See* Species specificity (uniqueness) of traits
Unity of humankind vs. racism. *See under* Racism
Universality of human nature traits
  as species-widely distributed vs. racism, 66
  vs. typicality (*see under* Typicality)

Vagueness
  and evolvedness of traits, 153–154
  and graded dehumanization, 235–236
  and species boundaries, 6, 37, 96–97
Value-ladenness of science, 97–98, 208–210, 238–241
Variation. See also *Differences*
  in body height, 51, 74–75, 158–160
  essentialist discounting of (*see under* Capacities of individuals)
  and membership conditions, 45–49, 127–129, 179–180
  and normalcy, 16, 19–20
  and population thinking, 33, 38–40, 49–57, 63, 126–127, 176–178, 179, 185
  and typicality of traits, 132–134, 137, 160–166, 224–225
Vernacular concept of human nature, 16–18, 31–32
  and dehumanization, 16–18
  and functional perspective, 29

vs. scientific concept of human
    nature, 32, 87
as socially perspectival, 16, 28–32, 87
Vitamin C, 173–174

Walsh, Denis, 49, 57, 169–170, 174–179
Waters, Kenneth C., 194, 227
Weber, Marcel, 260n7
Weismann, August, 62, 64–67, 80
West-Eberhard, Mary J., 68, 149
Western, educated, industrialized, rich,
    and developed countries (WEIRD),
    123
Williams, Raymond, 1
Wilson, Robert A., 94
Wittgenstein, Ludwig, 3, 239, 241

www.ingramcontent.com/pod-product-compliance
Lightning Source LLC
Chambersburg PA
CBHW021343300426
44114CB00012B/1060